Surrogate Motherhood Families

Olga B.A. van den Akker

Surrogate Motherhood Families

palgrave
macmillan

Olga B.A. van den Akker
Psychology
Middlesex University
London, United Kingdom

ISBN 978-3-319-60452-7   ISBN 978-3-319-60453-4   (eBook)
DOI 10.1007/978-3-319-60453-4

Library of Congress Control Number: 2017947072

Cover illustration: Photoshkolnik/getty images and Vlady9/getty images

Printed on acid-free paper

This Palgrave Macmillan imprint is published by Springer Nature
The registered company is Springer International Publishing AG
The registered company address is: Gewerbestrasse 11, 6330 Cham, Switzerland

I dedicate this monograph to all those who have made my work on reproductive health psychology possible.

Preface

This comprehensive monograph covers the research, theory, policy and practice contexts in which surrogate motherhood arrangements are carried out, and how the parties involved adapt or fail to adapt to continuing changes in public opinions, advances in technologies and new legislations. All chapters explain specific aspects of surrogacy in depth using some, but by no means all, of the available research evidence. From the first few cases of surrogacy published in the media in the 1980s continuing through the millennium right up until today, the focus has been on the controversies surrounding this practice rather than the harmony it has brought to many lives. Since the popular media have been the main source of information on surrogacy in the United Kingdom, population perceptions of surrogate motherhood arrangements have not generally been favourable. Despite the negative press there has been a sharp rise in surrogate motherhood arrangements. The book draws heavily upon the author's own psychosocial research on surrogate motherhood generously funded in the 1980s and 1990s by the National Health Service Research and Development. The book chapters are organised into three parts and have been divided into distinct but overlapping areas so that each can be read on its own or as a whole.

The monograph starts with a general introduction to surrogate motherhood in Part 1, which explains in detail what it encompasses and how it works (Chap. 1); the relevant theories and population attitudes shaped largely by the media (Chap. 2) and the practice and organisational and professional aspects of different surrogate motherhood processes (Chap. 3). The second part covers chapters on surrogate mothers (Chap. 4); fathers and/or mothers commissioning a surrogate baby (Chap. 5); the implications of separation and parenting (Chap. 6) and research on the surrogate baby who grows up to be an adult conceived via disparate surrogate motherhood models (Chap. 7). The final part (Part 3) outlines the particular psychological/emotional, social, financial, medical, ethical and legal costs and benefits associated with cross-border arrangements (Chap. 8); global ethical, moral and human rights issues (Chap. 9) and the development of and influences upon British legislation (Chap. 10). Within each of these three parts, in-depth accounts of psychosocial functioning through quantitative and qualitative research as well as theoretical perspectives are provided. The book is pitched for a wide readership focusing on academics, professionals/practitioners and students interested in alternative forms of parenthood. The monograph demonstrates the complexities, competing interests and interactions between legal, organisational, personal, social, cultural and biological/genetic parenthood.

London, UK   Olga B.A. van den Akker

Contents

Part 1 Surrogate Motherhood in Historical, Theoretical, Research and Organisational Contexts   1

1   Introduction to Surrogacy: Historical and Present Day Context   5

2   Theoretical Perspectives and the Social Context   39

3   The Process of Surrogacy   59

Part 2 Research on Surrogate Mothers, Commissioning Parents, Parenting and the Surrogate Offspring   75

4   Surrogate Mothers   79

5   Commissioning Parents   119

6   Separation and Parenting a Surrogate Baby   147

7   Individuals Born from Surrogate Mothers   169

Part 3   Cross Border Surrogacy, Ethical, Moral,
Human Rights Contexts and Legal Frameworks   195

8   Cross-Border Surrogacy   199

9   Ethical, Moral and Human Rights Considerations
     in Surrogate Motherhood   231

10   The Legal Framework   269

Abbreviations   307

Index   311

List of Figures

Fig. 1.1 Showing process and outcomes in gestational surrogacy
          for (a) the surrogate and (b) the intended recipient(s)   12
Fig. 1.2 Showing process and outcomes in genetic surrogacy
          for (a) the surrogate and (b) the intended recipient(s)   13
Fig. 2.1 Principles underlying feminist theories opposing surrogate
          motherhood practices   43
Fig. 4.1 Concerns about not being able to cope with their
          own emotions   94
Fig. 4.2 Concerns about the feelings of their own children   94
Fig. 5.1 Showing the main aetiologies for repeated implantation
          failure (Adapted from Margalioth et al. 2006)   122
Fig. 5.2 Showing genetic link possibilities in the surrogate baby   126
Fig. 5.3 Showing disclosure of IVF, sperm or egg donation or adoption 133
Fig. 5.4 Showing non-disclosure of IVF, sperm or egg donation
          or adoption   133
Fig. 6.1 Basic principles of attachment theory   155
Fig. 6.2 Showing the importance of communication and
          responsiveness to attachment formation   156
Fig. 6.3 Showing the importance of maternal sensitivity to attachment 157
Fig. 7.1 Individual and collective responsibility needs to be taken
          in research, policy and clinical practice as shown in the
          Prevention, Outcomes, Consequences (POC) model
          (Adapted from van den Akker 2013)   170

Fig. 8.1   Problems identified in cross-border surrogacy using
           surrogates from developing nations   202
Fig. 9.1   Showing the four basic ethical principles   237
Fig. 9.2   Showing religions with known moral objections/permissions
           of surrogacy   238
Fig. 9.3   The main ethical problems identified in surrogate
           motherhood   239
Fig. 9.4   Showing the increasing number of authorities responsible
           for bringing about and placing a surrogate baby, compared
           to adoption and traditional conceptions   241
Fig. 9.5   NHS costs of single, twin and multiple births
           (Adapted from Ledger et al. 2006)   245
Fig. 9.6   Showing the five areas of oppression as identified by
           Young (1990)   251
Fig. 9.7   A monitoring and registration system for surrogate births   256
Fig. 9.8   Some of the medical, social and psychological risks identified 257

List of Tables

Table 1.1   Terms used to distinguish the two broad types of surrogacy   17
Table 1.2   Some celebrities using surrogacy (age at the time of
             surrogacy, if known)   26
Table 3.1   Showing an example of a surrogacy 'package' deal   66
Table 6.1   Showing universally likely, possible and unlikely parental
             rights and responsibilities following surrogate motherhood
             arrangements   148
Table 8.1   Listing a number of countries where commercial or
             altruistic surrogate motherhood arrangements are allowed
             or explicitly not allowed (up to January 2017)   212

Part 1

Surrogate Motherhood in Historical, Theoretical, Research and Organisational Contexts

A surrogate motherhood arrangement is used to help individuals who cannot achieve a successful pregnancy to term such as infertile hetero-sexual couples, same-sex couples or single men and older single women to achieve parenthood. Surrogate motherhood has been described as a disaggregated process, where different components are mobilised to achieve the commissioned outcome. Chapter 1 outlines in detail what the process, terminology and historical context of surrogate motherhood mean to the individuals within the triads and to the society in which it takes place. The process is seen as becoming increasingly commodified. It is unusual in reproductive terms because in part it reflects a business transaction, but also a transaction based upon faith and trust; it is emotionally and financially burdensome and involves an inherent power and inequality component. Despite this, it is increasingly gaining in popularity; is mistakenly heralded as a close approximation to traditional conception; and some parents are content to deny any difference brought about through the disaggregated process to gain their baby.

Research described throughout this book shows that in successful cases, the surrogate mother tends not to have concerns or emotional pain upon relinquishment of the baby, not even if it is genetically related to her—much like her other children. The baby, child and adult conceived via this disconnected process may wish to understand it, but may find himself/herself unable to locate and understand the actions of the gamete or embryo donor(s), the surrogate mother or the contract specifying the requirements of the arrangement. This contract may include the potential requirement for foetal reduction, abortion for foetal abnormalities or multiple births, a clause not to 'take' the baby with abnormalities or ill health detected at or after birth; the minimum and maximum amounts of money agreed for the arrangement and at what stage in the process this is paid and any other contractual specifications, including specific behaviours (in extreme cases this may include a surrogate mother not having sex with her husband/partner, or leaving her own children for the duration of the pregnancy). This is not similar to traditional families' experiences as they do not plan for these specifics which are devoid of emotion, warmth and compassion when they anticipate the arrival of a baby within the family—whatever this child may be. A grown-up child too will know—if told—this was an unusual conception, and may need to adapt and adjust to this knowledge and to the presumed expectations of the parents.

Also within Part I, theoretical perspectives on surrogate motherhood arrangements are described in Chap. 2, and the dearth of research testing these theories is notable. Chapter 3 outlines the organisational processes involved in surrogate motherhood arrangements and demonstrates that universally assisted conception using third parties (assisted reproductive technologies themselves, surrogates, gametes and embryos) is largely available to commissioning parents who can afford it, leading to social inequality within infertile populations. Inequality between commissioning parents and the usually less affluent or poor gamete or embryo donors and surrogate mothers is also evident. Furthermore, people in high- and middle-income nations are more able to access these services than people living in low-income countries, and within nations, local policies, religious and cultural values also contribute to determinants of access to assisted conception services, leaving many people to seek these services cheaper abroad. Surrogate motherhood arrangements therefore allow for family building with substantial differences from traditional families: they are unusual. These differences are among the foci of this monograph.

Later chapters in Part II detail the research on surrogate mothers, commissioning parents, separation and parenting and the offspring, and chapters in Part III focus on research, theory, policy and practice from different jurisdictions, depending upon where the research has been carried out. To date, most research emanates from a limited number of Western countries. However, where relevant, other countries where the research has taken place will be highlighted, since population attitudes, medical processes and government policies will differ. The availability of much information on the internet and via print and televised/radio media has made surrogacy practices and malpractices available to all who want to know about it. Such a diverse range of information from research, policy, legislation, diverse international media and practices includes surrogacy uses for convenience (Warnock 1984; Chliaoutakis et al. 2002); what is and is not legal (in the United Kingdom, see for example HFEA Act 1990; Brazier et al. 1998); and medically (BMA 1996) or socially desirable surrogacy (Appleton 2001). This book provides a systematic and comprehensive account of surrogacy research, theory, policy and practices. It does not cover the reasons or fate of those who do not seek surrogacy to overcome involuntary childlessness, except to illustrate related points. Many infertile people never seek treatment (Greil and Mcquillan 2004), many of them are too depressed to seek help (Crawford et al. 2017) and they are not represented here. The majority of the chapters also provide an overview with historical and present-day perspectives, and, where possible, changes in practices over time are discussed within their respective social and cultural milieu.

References

Appleton, T. (2001). Surrogacy. *Current Obstetrics & Gynaecology, 11*, 256–257.

Brazier, M., Campbell, A., & Golombok, S. (1998). *Surrogacy review for health ministers of current arrangements for payments and regulation*. Report of the review team. Cm 4068. London: Department of Health.

British Medical Association. (1996). *Changing conceptions of motherhood. The practice of surrogacy in Britain*. London: British Medical Association Publications.

Chliaoutakis, J., Koukouli, S., & Papadakaki, M. (2002). Using attitudinal indicators to explain the public's intention to have recourse to embryo and surrogacy. *Human Reproduction, 17*(11), 2995–3002.

Crawford, N., Hoff, H., & Mersereau, J. (2017). Infertile women who screen positive for depression are less likely to initiate fertility treatments. *Human Reproduction, 32*(3), 582–587.

Greil, A., & Mcquillan, J. (2004). Help seeking patterns among subfecund women. *Journal of Reproductive and Infant Psychology, 22*(4), 305–319.

Human Fertilisation and Embryology Act. (1990). London: Her Majesty's Stationary Office.

Warnock, M. (1984). *A question of life: The Warnock report on Human fertility and embryology.* Oxford: Blackwell.

1

Introduction to Surrogacy: Historical and Present Day Context

The historical context and terminology used to fit unusual reproduction via surrogate motherhood into traditional models of the family are introduced in this chapter. Conflicting definitions are highlighted and attempts to minimise difference in favour of poorly working realities are shown not to be beneficial in the long term. Family differences have a legitimate place in society as evidenced by innovative fertility treatment options and solo, same-sex and older peoples' lifestyle choices. The different types of surrogacy, reasons for using surrogacy and the major issues for those involved are introduced. Inequalities, human rights, exploitation and the commodification of children are also shown to exist. Although most of the research is concerned with psychosocial issues, this chapter also draws on research from different disciplines.

Although genetic surrogacy is reported to have a history in ancient times (Schenker 1997), there is no acceptable precedent in purposefully conceiving, contractually gestating and anticipating relinquishing babies. Surrogate motherhood therefore is an unusual form of reproduction. The context of family building such as single, same-sex, unmarried, divorced, remarried, partnered, adoptive and reproductively assisted families common today provides the backdrop to surrogate build families. All of these family forms bear little resemblance to what is generically referred

© The Author(s) 2017
Olga B.A. van den Akker, *Surrogate Motherhood Families*,
DOI 10.1007/978-3-319-60453-4_1

to throughout this book as 'traditional' nuclear families (acknowledging there is no such thing as only one type of family, only a common type within any culture that new ones are measured against), yet the comparisons are consistently made even by those embarking upon these alternative family formats. Since diversity in family building is now commonplace, permanent and no longer alternative (Bernardes 1993; van den Akker 2012), the 'difference' of surrogate build families is just one of many differently formed family structures (van den Akker 2016). Surrogate motherhood is nevertheless unusual in reproductive terms. Third-party-assisted conception family discourses continue to be homogenised; emphasising resemblance even when there is none. Third-party input such as donated gametes/embryos or a surrogate mother's gestation of the commissioning parents' baby is marginalised or minimalised. Even in research and theory, kin relationships in third-party-assisted conception are generally understudied (Carsten 2004; Millbank et al. 2016) and little is known about the new biogenetic relationships which are choreographed in society within third-party families (Thompson 2005). According to Levine (2008), non-traditional families use traditional and radical biogenetic reference points, depending upon their disposition. Others referred to this as cognitive dissonance, which is evident in many different types of third-party-assisted conception families, including surrogates and commissioning parents and those associated with them (van den Akker 2007). Behaving in a way that may go against individuals' beliefs may exacerbate the fragility and uncertainty of a complicated and fragmented route to building a family.

Background

Surrogate motherhood is probably the best example of a complex and reproductively unusual arrangement separating maternity from social motherhood. The progress in reproductive technology raises many social, psychological, medical and legal issues (van den Akker 1998). These issues need to be understood within the sociocultural context in which this takes place. Psychological factors need to be studied and acknowledged to determine the palatability of surrogacy as it takes place in the United Kingdom and across the world today. Public opinion does not always favour surrogacy as a suitable option for infertile women unable to carry a child to term themselves or for men. Several enquiries and committees have been set up over the years, consisting of a diversity of distinguished individuals discussing the need for or against legislation within the procreative arrangement between three or more parties (see also Chap. 10). Other influential bodies, such as the British Medical Association (BMA 1996), acknowledged the inevitable practice of surrogacy and issued new guidelines for good practice and support to those involved. However, despite the documentation provided by these learned bodies, little information has been available on surrogacy, particularly evidence-based information concerning the long-term effects on intended (the mother or father commissioning) and surrogate mothers (the mother carrying and delivering the baby), and on the long-term effects of these diverse arrangements upon the resultant offspring and wider family network. Blyth's (1994, 1995) and van den Akker's (1998, 1999, 2000) interest in surrogate motherhood was fuelled by what was at a time a practice operating in a moral, ethical and legal vacuum. van den Akker applied for and gained a substantial Senior Research Fellowship grant from the National Health Service (NHS) Research and Development Office to carry out a programme of studies over a five-year period on this topic. Some of the progress which has been made in the last three decades to systematically study the issues involved has therefore come from these research efforts which have been used abundantly in this monograph.

Research and other media commentaries reporting on surrogate motherhood from a moral, ethical, legal, philosophical, social and psychological perspective either condone or condemn the practice. Both condoning and condemning stances are argued from opposing feminist, ethical, human rights and legal premises. All perspectives agree that selective 'inbreeding' and the exploitation and commodification of baby buying is unacceptable. These fears are well justified, and these too are reported in this book. However, the social context which demanded the need for surrogacy has only too often been ignored. This book tackles these issues to clarify the costs and benefits of surrogate motherhood and to address variables which should be clarified with appropriate future research. One important consideration in assessing the appropriateness of the legal, sociocultural, moral and psychological involvement in surrogate motherhood is one which relates to the concepts of reproductive and sexual rights. No one can even try to understand the meaning these rights have for different people. These rights should be interpreted within the context of the individual's age, marital status, economic and social conditions, religious and ethical identity and the stage in their life cycle. Thus, reproductive and sexual rights may not have universal applicability; they can shift in importance over the lifespan. There is also concern about the assumption of rights. If these rights refer to human liberty, then they are substantial. If these rights concern the satisfaction for the need of something wanted (such as a baby in commissioning parents' cases, and money in surrogate mother cases), these rights may have to be assessed. Addressing the rights of the commissioned offspring who may not know that donated gametes and/or surrogates, contracts and money were used in their conception may prove even more difficult. Provided no harm is done, or no intention to cause harm is the principle underlying these rights, they too may take on the acceptability society strives for.

Historical Perspective

It is estimated that two-thirds of Britain's pre-millennium surrogate babies were conceived using genetic surrogacy. This form of surrogacy is less time- and money-consuming and has higher success rates than gestational surrogacy, but carries with it some risks unique to this option. Firstly, a surrogate mother can be infected through the insemination of the commissioning husband's sperm. This is particularly risky if the insemination takes place outside the involvement of a clinic licensed to carry out such procedures. Secondly, there may be a greater risk of emotional distress for the surrogate mother who gives up her genetically linked baby, and for the commissioning mother who accepts a baby which is not hers genetically. This risk of greater emotional distress may also extend to the child who has a closer early link with the surrogate mother, both gestationally and genetically than with the commissioning parent fulfilling the social link. It is impossible to speculate on the relative importance of the gestational, genetic or social bonds between the parents and children, but some reports tend to indicate that knowledge of the people involved;

that is, 'openness' is less likely to have emotionally adverse consequences than anonymous arrangements between the parties involved. The effects on the surrogate and commissioning mother of genetic surrogacy will be discussed in more detail in Part II.

Gestational surrogacy, which became increasingly popular in the latter part of the 1990s, is not without its problems either. Here, the procedure involves a range of medical and health care involvement as well as a greater financial and technological burden. Gestational surrogacy has 'low' success rates and can be immensely costly. There are instances where infertile couples will choose the gestational surrogacy option first and failing that, after many attempts to establish a pregnancy in the surrogate, will opt for a genetic surrogacy instead. This option is far more regulated and controlled than genetic surrogacy, because of the clinical involvement. Clinics are legally bound to comply with strict regulation imposed by regulatory bodies. Counselling and ethical committee involvement is mandatory, as is testing for infectious diseases or genetic conditions.

In the United Kingdom, Kim Cotton was the first publicly talked-about surrogate mother in 1985. She has since set up and run an agency specifically to introduce infertile couples to women willing to carry a baby for them, the surrogates. Cotton has been a prolific advocate of both forms of surrogacy, until the publication of the Brazier report. It was in part the dissatisfaction with the conclusions of the report that finally made Cotton resign as chairperson of Childlessness Overcome Through Surrogacy (COTS), and to announce the problems she herself encountered during her first genetic surrogacy experience. Kim Cotton became widely known as the woman who gave up a genetically related child to an unknown couple in the United States. She gave her baby up for money (£6500) and did not know the couple who were to become the parents of her child, suggesting she was unlikely to have done this for altruistic reasons. The arrangement was brokered by an American company, although Kim carried and delivered the baby in the United Kingdom. Since those 1980s pioneering surrogate motherhood arrangements in the United Kingdom, surrogacy has become increasingly recognised as an option for infertile couples, although it remains an unpalatable option for many people as is demonstrated by the UK media (van den Akker et al. 2016a, b). Despite the efforts to operationalise UK surrogacy, the controversy surrounding it remains as intense as ever. This monograph explores the many issues identified across all aspects of past and present surrogate motherhood arrangements, focusing on the United Kingdom, but also covering research from across the world as examples influencing local practices. The future of surrogacy in the context of current legislation and population practices, needs and opinions remains controversial, but new and targeted research-informed policy and practice could move surrogate motherhood arrangements to a more harmonious place in modern society alongside other openly less traditional family forms.

However, like the early reports of donor insemination (DI), followed by in vitro fertilisation (IVF) and oocyte donation, controversy was replaced by acceptability over time (Addelson 1990). With surrogacy, acceptability has been slow to emerge probably because 'parenthood' here is linked to the legal position attributing motherhood (in the United Kingdom) to the surrogate birth mother. In other forms of third-party conception such as gamete or embryo donation, the 'route' (origins of mitochondria/gametes/embryos from a third party) has no legal connotations. In this way, the medical interventions necessary in third-party reproduction (mitochondrial/gamete/embryo donation) may define parenthood as social parenthood, but in surrogacy there is also a legal definition: a birth mother. Traditional family theory refers to traditional (facilitated by nature, heterosexual, biological) conception and parenthood is thus defined in biological terms (Bartholet 1993; Bartholet et al. 1994). The difference from traditional families created in all third-party reproduction, including mitochondrial, sperm, oocyte, embryo donation and surrogacy, tends to be hidden, marginalised or denied (van den Akker 2001, 2007, 2016). When this happens, a new reality or script is created (Strathern 2002), leaving the resultant children potentially with inaccurate family histories.

Surrogacy

In a review of gestational surrogacy, Brinsden (2003) said 'it is not the treatment of the parties involved in (gestational) surrogacy that is complicated, but the preparation of them, with the proper provision of advice:

legal and medical, the proper provision of counselling and the careful selection of a suitable host'. This statement reflects the interacting aspects of surrogate motherhood and demonstrates how one cannot be seen outside of the context of the other aspects. Over a decade later, it is fair to say that the preparation of the intended individual(s) remains complicated as will be shown. Surrogacy is an arrangement where the traditional motherhood/parenthood functions are fragmented. In surrogate mother arrangements, one woman—the surrogate mother—becomes pregnant through assisted reproductive techniques (ART) or insemination, carries and then delivers a baby for another, usually infertile woman, or for a man or couple who cannot achieve a pregnancy—the recipient or commissioning couple. The baby is handed over to the commissioning recipient(s) immediately or soon after birth (Appleton 2001; Sharma 2006), who then raise it as their own. Two types of surrogacy are widely practised: gestational surrogacy and genetic surrogacy. There are a number of options within the two broad types of genetic and gestational surrogacy.

Gestational Surrogacy

Gestational surrogacy is the more complex, more time consuming and more expensive option, requiring fertility clinic involvement and the creation of an embryo ex vivo (outside the uterus) using IVF (in glass) (ACOG 2008). This process requires medical intervention and the resultant child could not exist without the explicit selection of gametes, the IVF process and the embryo transfer (ET). A surrogate is not genetically related to this foetus/child. In this arrangement, in most cases the recipient(s) are entirely genetically related to this child, the recipient mothers' oocytes (eggs) and the recipient fathers' sperm. However, donor eggs, sperm or donor embryos may also be used. The donated gametes (oocytes or sperm) or embryos can be left over from other patients, or donated by altruistic or commercially paid donors. In all these cases, the surrogate remains genetically unrelated to the child and the recipient mother, the father or both are genetically related, or neither recipient is genetically related to this resultant child. The new intended couple were not involved in its gestation. The intended commissioning recipient(s)

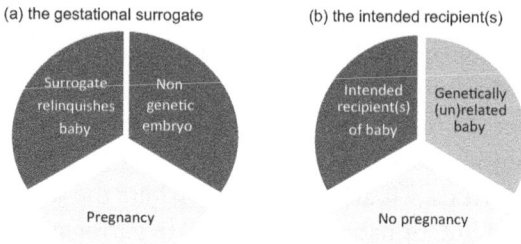

Fig. 1.1 Showing process and outcomes in gestational surrogacy for **(a)** the surrogate and **(b)** the intended recipient(s)

in gestational surrogacy is/are in effect three steps away from traditional conception; (1) they miss out on sex to procreate; (2) on the nine-month gestation and (3) the subsequent labour and delivery of the baby. In addition, when the commissioning parents contribute the embryo they require oocyte retrieval and IVF to fertilise the egg. After this long and complicated process involving third parties, they become the recipients of a newborn baby. The surrogate also finds herself involved in a medical and technological conception, as far removed from natural conception as possible, and is in effect also just three steps removed from traditional childbearing because; (1) she does not conceive traditionally; (2) the baby is not genetically linked to her and (3) she relinquishes a baby she gestated. Figure 1.1a, b are illustrations of the fragmentation of functions, assuming the traditional mater semper est principle consists of one whole circle, that is, the 'mother(hood) is always certain' principle is composed of a woman, a pregnancy and her genetic embryo/baby.

Genetic Surrogacy

Genetic or 'traditional' surrogacy involves a process that mimics natural conception. Here the surrogate mother is artificially inseminated with the intended father's (or donor) sperm via intrauterine insemination (IUI) at home or in a clinic, although it can be done much more expensively via IVF to fertilise her egg. She gestates a baby similar to her own children and after birth she gives away a child bearing her own genetic material to the commissioning recipient(s) (ACOG 2008). The commissioning recipient(s) here take home a baby which is only partially or not at all genetically related to them, and was not gestated by them. Here the recipient(s) couple or individual is only one step away from adoption but far removed from traditional conception, where neither a full genetic link nor gestation was part of the family building process. However, for the surrogate, it is as close to a traditional conception, gestation and delivery as if this was a pregnancy planned for her own family building purposes. The child in this case is still genetically related to the surrogate, but is genetically unrelated to one or both of the recipient(s), who may, depending on which country the arrangement takes place, need to go through a legal process to adopt the child as theirs or seek parental responsibility once it is born (Fig. 1.2).

Either type of surrogacy can be used by people for medical or social reasons. In practice, social surrogacy usually relies on the availability of commercial surrogates; whereas non-commercial or altruistic surrogacy tends to be used in cases where there is a medical need to use a surrogate, provided it is available and legal in the country where it is sought. The different types of surrogate mother will be discussed further in a later chapter (Chap. 4). The expectations of women who become surrogates—for recipient(s) who have either social or medical reasons to use surrogacy—are likely to be different. However, little research (see, e.g. van den Akker 2005, 2007) has clearly differentiated between these two types of surrogate.

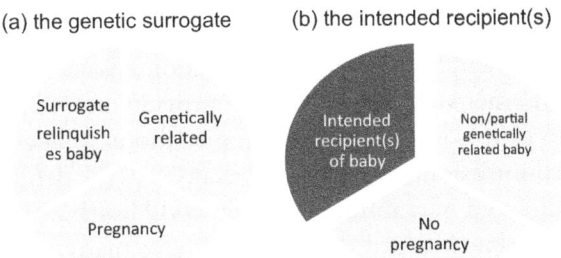

(a) the genetic surrogate   (b) the intended recipient(s)

Surrogate relinquishes baby — Genetically related

Intended recipient(s) of baby — Non/partial genetically related baby

Pregnancy   No pregnancy

Fig. 1.2 Showing process and outcomes in genetic surrogacy for (**a**) the surrogate and (**b**) the intended recipient(s)

Commissioning Parents

Commissioning parents are individuals or couples who seek to build their family using a surrogate mother to carry a baby for them. They have been defined as contracting, intended, commissioning and social parents (Ciccarelli and Beckman 2005) depending on the stage they are at. These terms themselves frame the parents in a way that explicitly connotes contracts, an 'intent' to enter an agreement and a financial transaction. If the parents are genetically related to the surrogate baby, referring to them as the 'social' parents leaves out some important information they share with the baby, a difference not present in traditional social parents in adoption. However, for the purposes of the chapters presented here, the term 'commissioning parents' will be used throughout, as it is not possible to always know with certainty that the terminology used in the research compiled for this monograph accurately reflected the stage at which the parents to be were when the specific terms were used. Second to that reason, the majority of research, theory and practice use the reference term 'commissioning parent(s)'.

According to the Human Fertilization and Embryology Authority (HFEA; HFEA 2009, S1, 1.3), no real definitions for commissioning parents are provided, but an explanation is noted; commissioning couples 'may both be the genetic parents, or just one or neither of them may be genetically related to the child. This will depend upon the sources of the egg and sperm used to create the embryo'. In some instances it may be that both the sperm and the egg were provided by donors, so that the resulting child is not related genetically to either the surrogate mother or the commissioning couple. In such cases the commissioning couple cannot apply for a parental order (PO)—see section below.

Accessible internet websites, which can provide useful knowledge and information to people interested in the topic, also pass judgement on the definition of commissioning mother, for example; 'The definition is quite to the point and void of emotion' (Women and Health 2016), but at the same time it is acknowledged that 'she will have many important emotions during the surrogate arrangements and beyond.' The sparse interest in the definition of the commissioning parents demeans their importance in the arrangement. If it was not for the commissioning parents' need to use surrogacy, the surrogates would not have a role. Also, the commissioning parent(s), with a few notable exceptions, will take on the full lifetime responsibility to parent the resultant child, and will eventually be the legal parents of the surrogate baby.

Surrogate Mothers

Surrogate motherhood has been defined much more prolifically in a number of different ways. For example, the *Oxford Dictionary* defines a surrogate mother in two ways, as 'A person or animal which takes on all or part of the role of mother to another person or animal' and as 'a woman who bears a child on behalf of another woman, either from her own egg fertilized by the other woman's partner, or from the implantation in her womb of a fertilized egg from the other woman' (http://www. oxforddictionaries.com/definition/english/surrogate-mother; accessed 10 January 2016). According to the first definition, surrogate mother refers to the recipient or commissioning mother as the surrogate, the second the mother who carries and relinquishes the baby. The *Cambridge Dictionary* defines a surrogate mother as a woman who has a baby for another person who is unable to become pregnant or have a baby (http://dictionary. cambridge.org/dictionary/english/surrogate; accessed 17 January 2016).

A Wikipedia definition of surrogacy is much more elaborate, and as is commonly the case with Wikipedia definitions and it is selectively referenced and not necessarily accurate (accessed 17 January 2016). Wikipedia defines surrogate motherhood as 'A surrogacy arrangement or surrogacy agreement is the carrying of a pregnancy for intended parents.' There are two main types of surrogacy: gestational surrogacy (also known as host or full surrogacy which was first achieved in April 1986) and traditional surrogacy (also known as partial, genetic or straight surrogacy). Gestational surrogates are also referred to as gestational carriers. In traditional surrogacy, the surrogate is impregnated naturally or artificially, but the resulting child is genetically related to the surrogate. In the United States, gestational surrogacy is more common than traditional surrogacy and is considered legally less complex (Gaither 2015). Intended parents may seek a surrogacy arrangement when either pregnancy is medically impossible, pregnancy risks present an unacceptable danger to the mother's health or is a same-sex couple's preferred method of procreation. Monetary compensation may or may not be involved in these arrangements. If the surrogate receives compensation beyond reimbursement of medical and other reasonable expenses, the arrangement is considered commercial surrogacy; otherwise, it is referred to as altruistic. The legality and costs of surrogacy vary widely between jurisdictions, sometimes resulting in interstate or international surrogacy arrangements (https:// en.wikipedia.org/wiki/Surrogacy; accessed 17 January 2016). Wikipedia also cites the following as the historical perspective on surrogate motherhood: 'Having another woman bear a child for a couple to raise, usually with the male half of the couple as the genetic father, is referred to in antiquity. Babylonian law and custom allowed this practice, and an infertile woman could use the practice to avoid a divorce, which would otherwise be inevitable (Postgate 1992).'

The HFEA in the United Kingdom is an independent regulator overseeing the use of gametes and embryos in fertility treatment and research. It is accessed by clinicians, researchers and the public and provides a considerably less detailed description of surrogacy; 'Surrogacy is when another woman carries and gives birth to a baby for the couple who want to have a child. The HFEA does not regulate surrogacy.' They cautiously add, 'We recommend that you should seek legal advice before proceeding with this option' (http://www.hfea.gov.uk/fertility-treatment-options-surrogacy.html; accessed 17 January 2016). Most other countries use similar definitions. In the United States, the American Society of Reproductive Medicine (ASRM), for example, distinguishes between a gestational carrier (a woman who carries a pregnancy for another—in the United Kingdom a gestational surrogate) and a surrogate (a *woman who* donates her egg *and carries* a pregnancy for another), in the United Kingdom a genetic surrogate. The ASRM defines surrogacy as 'an arrangement in which a woman is inseminated with the sperm of a man who is not her partner in order to conceive and carry a child to be reared by the biologic (genetic) father and his partner. The surrogate is genetically related to the child. The biological father and his partner usually must adopt the child after its birth' (https://www.asrm.org/Templates/ SearchResults.aspx?q=surrogacy&p=1, accessed 19 January 2016). They define a gestational surrogate as a 'Gestational Carrier' who is 'a woman who agrees to have a couple's fertilized egg (embryo) implanted in her uterus. The gestational carrier carries the pregnancy for the couple, who usually has to adopt the child. The carrier does not provide the egg and is therefore not biologically (genetically) related to the child.' Referring to a surrogate mother in these non-maternal terms, 'gestational carrier', removes the maternal and personhood from the terminology, therefore rendering the arrangement a more clinical, rather than maternal, transaction. However, the surrogate mother has also been called the real or actual mother of the surrogate baby, as she gestates and delivers the baby. Viewed in these terms, giving the actual (the surrogate) mother, 'surrogate' status, demeans her role as mother (Hanafin 1999; Tangri and Kahn 1993) and delegitimises her rights to a continuing relationship with the baby (Jaggar 1994, p. 379).

Table 1.1 lists some of the common terms used to define the two types of surrogate/surrogacy. The terminology and definitions used therefore frame the concept differently (Beeson et al. 2014). If, for example, the 'maternal' or 'motherhood' is removed from the surrogate motherhood transaction, it puts her mothering role into question, and suggests that surrogacy should be considered employment rather than an act of compassion, and surrogate children as 'products' or 'goods' (Darnovsky and Beeson 2014). It also allows for employment rights of the workers (the gestational carriers) and an evaluation of the product by the buyers. It may be that this marks the future of designer children. The business transaction analogy is in fact not at all new. The sale of gametes and embryos across different countries is already part of the process. A Google keyword search of 'embryos for sale' brings up pages of adverts for good-quality embryos (see for example the California Conceptions Donor Embryo Program). This example had some colourful headlines explaining the

Table 1.1 Terms used to distinguish the two broad types of surrogacy

Genetic surrogate	Gestational surrogate
Traditional surrogate	Host surrogacy
Full surrogacy	Partial surrogacy
Surrogate	Gestational carrier
	Global surrogacy arrangement

ease with which you can be helped to become a picture postcard family, cheaply and 'with a guarantee option available to qualified individuals'. The same is available for sperm and oocytes.

Altruistic Surrogacy

In altruistic surrogacy, the surrogate will not receive a fee for her services as a surrogate carrying the baby. Instead she is likely to receive money in compensation for travel to and from clinics, additional childcare costs for her own children, extra vitamins, and maternity clothes and so on. However, there have been questions about the amount of money that is paid for compensation, since that can be a substantial amount ranging from anywhere around the £10,000 mark, give or take 5 k either side, depending upon the 'experience' of the surrogate. Altruistic surrogates are in relatively rare cases friends or family of the recipient(s). The United Kingdom allows altruistic surrogacy, and in the majority of arrangements the parties do not know each other and are put in contact with each other via surrogate agencies.

Commercial Surrogacy

Commercial surrogates tend to be recruited into surrogacy agencies for money. These surrogates expect reimbursement for compensation in the same way as altruistic surrogates, but in addition expect earnings for loss of work or simply because they see these services as a job. With the addition of financial pay for services, most are not previously known to the recipients and some may live in other countries. These surrogates (depending on the country) can advertise their services or can be recruited and the process can be run as a business transaction. Across the world, large organisations and clinics run such businesses. There are fears about the ethical and moral rights and wrongs about these financial practices, with increasing reports of exploitation, baby farming, buying (specific) babies and targeted breeding opportunities circulating in the media and research reports. This is discussed in detail in Chap. 8.

Global Surrogacy

Contract pregnancies across different countries are always commercial and almost always gestational surrogate arrangements (Henaghan 2013). Here the financial transaction is very important, as international or cross-border arrangements are brokered by clinics and services are paid for following a contract, specifying what is and is not included.

Incidence and Practice Perspectives

From its rare beginnings in biblical times (Harrison 1990) surrogacy is now being carried out with increasing frequency. National statistics are difficult to obtain, and are known not to reflect the actual numbers using this means to build a family. The US Office of Technology Assessment reported approximately 600 babies in 1988 (OTA 1988); Ragone estimated about 100,000 in 1994. In 2003, the United States made it mandatory for the accurate reporting of surrogate births as separate from IVF births. According to Ciccarelli and Beckman (2005) the ASRM reported 129,000 IVF births between 1988 and 1999, and (between the years 1991 and 1999), 1600 of these were surrogate births. In the United Kingdom, accurate reports are even more difficult to obtain. The BMA (BMA 1996) noted that the incidence of surrogacy arrangements was impossible to ascertain because of the informal arrangements taking place, making any form of regulation difficult. van den Akker (1999) obtained surrogate birth figures from British surrogate agencies from the period 1995 to 1998, as 332 surrogate births; however, the General Register Offices (UK) figures for the same period were substantially less ($n = 166$), suggesting that only half of commissioning parents went on to obtain a parental order (for more detail on this see Crawshaw et al. 2013).

Chronologically, the first mention of a surrogacy case was in 1954 (see Schmulker and Algen 1989) followed by a report of the first American surrogate birth in 1985 (Utian et al. 1985) where genetic and gestational surrogacy is practised and commercial arrangements are accepted in some states. The United Kingdom saw its first surrogate in 1985 recruited through an American organisation to become a genetic surrogate for an American couple. The arrangement had devastating consequences for the surrogate (Cotton 1985) and led to the Surrogacy Arrangements Act (1985), for the purposes of limiting the practice. In the United Kingdom, unlike the United States, commercial surrogacy under the Surrogacy Arrangements Act is illegal, but altruistic surrogacy is allowed. The BMA (BMA 1990), after initially objecting to the practice, also conceded reluctantly that it would not be possible for medical professionals to ban the practice—if legislation stipulates it is legally allowed. For more information on this and subsequent reports and changes in legislation, see Chap. 10. At about the same time, the HFEA Act (HFEA Act 1990) was passed in Parliament and so surrogacy became legal as a practice in the United Kingdom. In 1990, the United Kingdom made it possible for recipient parents, then recognised only as heterosexual couples in stable marital relationships, to become the legal parents of surrogate-born children via the transfer of legal parenthood route. It was not until 2008 that the UK Government recognised the more diverse needs for family building of people in non-heterosexual relationships—such as gay and lesbian couples and single men and older single women—and changed its laws accordingly.

Clearly surrogate motherhood in the 1980s and 1990s was allowed, but not welcomed or facilitated as a means to overcome involuntary childlessness. The Surrogacy Arrangements Act (1985) stipulates the commissioning couple must find their own surrogate mother, and at the same time surrogates were not allowed to advertise their services as surrogates, leaving surrogacy available only to couples who knew a close friend or relative willing to carry a baby for them. COTS and other agencies (such as the SPC, Surrogate Parenting Center) were set up to help put people in touch with each other. In 1996, the BMA reported surrogacy was an acceptable last-resort option for women who could not carry a baby to term for medical reasons. A year after the second BMA notification on surrogacy in the United Kingdom, the Brazier report (1998) was published following a working party commissioned by UK Health Ministers after another disputed custody case involving surrogacy. Not all the recommendations of the review of the Act have been enacted. More information on the Act and the BMA recommendations is described in the chapters on ethical and legal issues (Chaps. 9 and 10 respectively).

As mentioned, the legislation also did not provide a legal opportunity for single men or women, or individuals in same-sex relationships to build their family (van den Akker 1999). That did not happen until the cultural context changed when same-sex couples were accepted as adoptive parents, as civil partners and finally as married couples that they could be considered for surrogacy over a decade into the millennium, but even then the first cases of same-sex and single parent surrogacy were commissioned abroad. Since the earliest reports of surrogacy in the United Kingdom, surrogate motherhood has been treated savagely by the British press (van den Akker et al. 2016a, b), and as a result of the public debates, the increasing visibility of surrogacy, and the fact that whatever the consequences, the reality was that infertile (and much later same-sex) couples ended up with surrogate babies was reported in the media. This fuelled the desires of many other involuntary childless couples to become parents of newborn babies. It also led the way for other women to make a difference to their own (via compensation or feelings of altruism) or others (the grateful recipients) quality of life. The agencies (COTS/SPC) were set up to deal with the demand and to pave a way for some sort of order in what could otherwise have turned into a largely invisible chaotic practice. After much to-ing and fro-ing, weighing up the anticipated uptake of surrogacy by UK couples, the consequences to surrogates, population attitudes (Poote and van den Akker 2009) and the very difficult decision to legalise parenthood via this third-party route, a decision was finally made five years later (see van den Akker 2002 for a more detailed context and historical account of British surrogacy policy and practice).

However, to date single men and women are still discriminated against, as they still cannot apply for a PO for a baby born to a surrogate mother. For example, recently (September 2015, Blackburn-Starza 2016a), the President of the Family Division refused a single man's application for a PO. The baby was conceived with his sperm, and he obtained an order in Minnesota which extinguished the surrogate mother's rights, establishing the baby's sole legal parentage with the father, and he was named on the birth certificate. The father took the baby to the United Kingdom.

Despite the Minnesota Law having removed the surrogate mother's legal rights to the baby, this did not apply to British law which does give her rights as the baby's parent. The father applied for a PO which, if granted, would extinguish the birth mother's legal rights in England and the father would be treated as the child's only parent. However s.54 (1 and 2) of the HFEA Act (HFEA 2008) only allows the Court to make a PO on the application of two people and for this reason the judge refused to grant a PO. The father then applied for a declaration that the legislation was in breach of the European Convention on Human Rights (see Blackburn-Starza 2016b). The Secretary of State for Health was invited to intervene and he conceded that on a very small point the legislation was incompatible with the European Convention on Human Rights. The judge made a declaration to this effect but he was still unable to make the PO. The law will change only when Parliament legislates on the point. It is unlikely that looking at a case like this, the welfare of the child or the two disparate 'parents' is in any way served well by the legislation. In a similar case, another single father of a surrogate-born baby adopted his baby to circumvent the child remaining a ward of court.

The increase in cross-border surrogacy, discussed in detail in Chap. 8, rendered the already compromised laws insufficient to deal with the range of arrangements people became embroiled in. So much so, that in November 2015, Hornsey, Smith, Norcross, Ghevaert and Jones published a 'Myth busting and Reform' report in an effort to push for a reform of the current surrogacy law (Hornsey et al. 2015). Their recommendations are varied and not necessarily in the best interests of the child, as will be shown later in the chapter on national surrogacy (Chap. 4). However, they, along with many others, have highlighted a need to refresh current legislation and practice. Discriminating against single parenthood is obviously antiquated in the West and does nothing to benefit society. Interestingly, for treatment using donated gametes (in non-surrogacy), the HFEA (2016) reveals that the NHS does not fund all gamete donor treatments, and women in same-sex partnerships and single women (or men) are even less likely to receive NHS funding for their treatment (HFEA 2016). Discrimination based upon sexual orientation and single parenthood therefore continues to be rife in assisted conception across economic and social/lifestyle strata.

Procreation and the Family

Having (wanting) children is an important milestone in most people's lives (Heinicke 1995), whether as a single person, in same-sex or heterosexual relationships, or when people are older in reproductive terms, parenting is desired across the world and across cultures (Inhorn 2003). Contextually however, there are subtle differences between and within cultures/nations. The traditional Western view of the family is as individualistic units sanctioned by the state, as marriages. Single parenting is less acceptable in individualist societies as social and financial (including tax relief) support for single parents is not equal (proportionally) to that of married couples, and as described above, although fertility treatment for single people is available, the NHS provision and legal provision for parenthood are selective. Nevertheless, many countries have now equalised same-sex with opposite-sex marriages and are moving to equalising family building opportunities, including the United Kingdom. Many non-Western countries have a more collective approach to health, wellbeing and the family. In some African countries, the collective approach includes the moving of children to other kinship networks, where they may stay for some or many years. The parenthood of the original parents is never questioned, and a child's return to the birth parents remains an option.

Having children is the most important part of investing in marriage in African (Ombelet et al. 2008) and other cultures and not being able to build that family is a significant failure in the eyes of many communities. Children are socioeconomic investments to look after their parents as they age (Dimka and Dein 2013). The drive to be traditional, to be seen to have children and the status associated with having a family leads to extreme measures in some communities. According to some African traditions, infertility and fertility are attributed to God's will or bad spirits and evil witches, and 'marabouts' (Muslim teachers with supernatural powers) and other traditional healers are the first resort for infertility treatments in some African countries. The consequences of not producing children can be harsh, for example:

- Infertility in women can be stigmatising, in, for example, the Gambia (Soundby 1997).

- In many African cultures, including South Africa, infertility is solely attributed to women and in such communities it is considered to be a taboo to discuss the fertility of men (e.g. Dyer et al. 2004, 2005, 2008).
- Some Rwandan infertile people, even in more recent times, are accused of witchcraft (Dhont et al. 2011).
- Men in South Africa may claim they have had children within other relationships (Dhont et al. 2010).
- Alternatively in Zimbabwe male infertility may be covered up by having close relatives impregnating the wife (Matetakufa 1998).

Although reasons for parenthood in developed countries tend to be individual rather than collective focused, (van Balen and Trimbos-Kemper 1995; Langridge et al. 2000; Cassidy and Sintrovani 2008; Purewal and van den Akker 2007), marital status is a less important reason for parenthood (Humphrey 1977), indicating differences between cultures. Nevertheless, male fertility is also associated with masculinity and female infertility with femininity in the West and these examples have led the way for continuing secrecy about infertility across the world with most infertile couples seeking treatments to overcome involuntary childlessness if they can. As a result of the huge social burden of infertility across the globe, assisted conception technologies are rapidly globalising to all pro-natalist developing societies, where children are also highly desired, parenthood is culturally mandatory, and childlessness socially unacceptable, with potentially severe consequences as noted above. The major forces fuelling the global demand for assisted conception include demographic and epidemiological factors, the fertility–infertility dialectic, problems in health care seeking, gendered suffering, and adoption restrictions (Inhorn 2003). The demand for overcoming involuntary childlessness including opting for infertility treatment therefore corresponds to the desires to have children (for whatever the reasons). However, the acceptability of some forms of treatment, particularly third-party treatment, also differs across the world. According to Bello et al. (2014), in Nigeria, the acceptability of IVF was 59.3%; 37.8% found surrogacy an acceptable alternative and 35.2% would consider egg and 24.7% sperm donation, confirming research in British populations (van den Akker 2001, 2007). The fact that egg and sperm donations are the least acceptable forms of ART highlights the importance of genetic links in many societies. Other studies report different percentages, but broadly the preferences follow the same pattern, with a full genetic and gestational link preferred, followed by no gestation and part to no genetic link (Lasker and Borg 1987; Krishnan 1994; van den Akker 2007), although there are exceptions (e.g. Miall 1989; Petitfils et al. 2017).

Reasons Why People Use Surrogacy

Medical and social reasons: Medical and social reasons to choose surrogacy as a means to build a family is explored in depth in Chaps. 3 and 5. Information is emerging on intended/commissioning couples and individuals seeking to use commercial or non-commercial altruistic surrogates. Single and gay men increasingly approach commercial surrogates to build their families. Elton John and his partner David Furnish, for example, used surrogacy because between them they cannot have genetically related or gestated children because they are gay men. They paid for a surrogate pregnancy, and depending on the company they used to broker these arrangements, they may have paid for specifically selected gametes (in their case oocytes from a surrogate or an oocyte donor). Ricky Martin is single (and gay so will be unlikely to ever have a relationship with a woman) and also used surrogacy to ensure he too had a chance to build a family. Unusually, this is one of the few commissioning recipients of a surrogate arrangement who was young. Most recipients of surrogate babies are older when they start, or end their family building through surrogacy. Robert de Niro (who already had children from previous relationships and had used surrogacy in another previous relationship), for example, and his current wife, Grace Hightower, were not of reproductive childbearing age when they commissioned a surrogate baby—she was 56 (Celebrity 2016). The fact that many women are older when they commission a surrogate baby is unsurprising because, although some simply left it too late, many others have attempted to conceive using IVF for many years before embarking upon surrogacy. Other women who do not have a sufficiently competent uterus to sustain a pregnancy are also known to use surrogates to fulfil their wish to become a parent.

Table 1.2 Some celebrities using surrogacy (age at the time of surrogacy, if known)

		Outcome
Robert de Niro (68)	Grace Hightower (56)	Singleton
Chris Bosh	Adrienne Bosh	Twins
Elton John (65)	David Furnish (50)	2 × singletons
Courtney B Vance	Angela Bassett	Twins
Keith Urban	Nicole Kidman (43)	Singleton
Kelsey Grammer	Camille Grammer	Two singletons
Ricky Martin (single)		Twins
Charlie Brooks (48)	Rebecca Brooks (43)	(Twin pregnancy) singleton survived
Matthew Broderick	Jessica Parker (44)	Twins
Chris Ivery (46)	Ellen Pompeo (44)	Singleton

Commodification Some celebrities have used surrogacy, for example, Jessica Parker in 2009 (she herself gave birth to a son seven years previously). She opted for surrogacy to have another child at the age of 44, and Nicole Kidman did so aged 43 (see Table 1.2). It is possible both also had difficulty conceiving because of their age, but we do know that both had tried to conceive for several years before trying surrogacy. 'Celebrity surrogacy' cases are readily picked up by the media (see e.g. van den Akker et al. 2016a, b), describing the practice as designer family building, without the inconvenience of bodily discomfort of pregnancy and the pain of giving birth at an age when this would have been possible. In reality, this seems rarely to be the case. Nevertheless, such exaggerated headlines and speculative media content can affect population attitudes to this practice; see for example: http://www.therichest.com/expensive-lifestyle/lifestyle/18-celebrities-who-used-surrogacy/

Reasons Why Women Become Surrogates

Altruism Altruism is believed to be a motivator for women to become surrogates in some cases. Altruism is seen as a motivational state in which increasing another person's welfare is the ultimate goal (Batson 2014) even if this is at the expense of the individual carrying out the altruistic act or behaviour. In altruistic surrogacy, one woman carries a baby for another person without expecting anything in return. The surrogate's sole aim is to selflessly help another person achieve parenthood (Poote and van den Akker 2009). Traditionally, the giving up of a baby is strongly discouraged as mother-child bonding is an important part of secure attachment formations in later life. In surrogacy, according to Cherry and Peppin (2005) a vice (the ability to detach from a child in utero) is turned into a virtue. Other reasons for becoming a surrogate include having the belief that they would be able to relinquish the baby after birth and based upon whether they knew and liked the commissioning couple (Hanafin 1984). This latter finding supported what Ciccarelli (1997) had mentioned previously, that humans have the ability to exercise control over their attachment needs. So, if people are given reasons not to attach or to detach, they are able to do that if they want to (Ciccarelli 1997).

Commodification Critics of surrogates focus on the commercialisation and potential commodification aspects of surrogacy. According to Brazier et al. (1998), commercial surrogacy is a lucrative business, and the more lucrative the job, the more women would get attracted to it. Indeed, Blyth (1994), Braverman and Corson (2002) and van den Akker (2007) reported that some women become 'altruistic' surrogates because of financial motivations, whereas others simply enjoy the pregnancy and childbirth, and others do it for a sense of self-worth and value. A proportion of surrogates are therefore women who have altruistically volunteered to become surrogates (Ragone 1994; Braverman and Corson 2002; van den Akker 2007).

Effects on Others in the Network

Family disapproval of surrogate motherhood has been evident in research which has asked the relevant questions. In van den Akker's studies (van den Akker 2001), surrogacy had a positive effect on close family members, and in Hohman and Hagan's (2001) study, the surrogate mothers' own children were not negatively affected by their mothers' involvement in a surrogate arrangement. However, Ciccarelli's (1997) research demonstrated less than a third of the extended family of surrogates were supportive, and many experienced interpersonal conflict, with over 40% losing a relationship as a consequence. Some surrogates also regret their involvement as surrogates (Blyth 1994; Ciccarelli 1997) and this may strengthen over time particularly when contact with the commissioning couple decreases over time (Ciccarelli 1997; van den Akker 2003, 2007). It is likely that adequate counselling (Ciccarelli 1997) over a long enough period of time (van den Akker 2001) is necessary to assist surrogates in particular after carrying, delivering and relinquishing the surrogate baby. Research into genetic surrogates' own children is insufficiently addressed and into the surrogate mothers' own parents is virtually non-existent. These relatives all lose a genetically related child to intended commissioning couples, with no opportunity to link and welcome the child into what they may perceive as the genealogic family.

With regard to commissioning parents' networks, these are generally supportive of the couples' resort to surrogacy to build a family (Ciccarelli and Beckman 2005), although many did not disclose beyond the immediate family. Embarrassment about finding out about a surrogate and revealing information about themselves were reported in Blyth's research (1995). Kleinpeter's (2002) study revealed half the families of the commissioning parents were supportive, and some had mixed reactions whereas friends were generally supportive.

Social Support in Surrogacy Arrangements

Research by Thorpe et al. (1992) demonstrated the benefits of social support during pregnancy and the detrimental effects of disrupted social support. The social support of surrogate mothers is likely to be disrupted by the increased attention they receive as a result of the arrangement. Whilst that is important, there is a danger that established social support is relinquished to some extent because the surrogate's networks are not as involved in the surrogacy process and the resulting child will not become a member of the family. From the intended family's viewpoint, the missing genetic link in partial surrogacy may pose a threat to the marital/family relationships as discussed in The Warnock Report (Warnock 1984). In addition, Golombok (1992) notes the effects reactions (such as being shunned) of family and local communities can have on surrogate and intended parents. There is little evidence of effectively delivered therapeutic interventions to ease the transition to parenthood via surrogacy, a significant shortcoming which will be addressed later in the book.

The Benefits of Personal Contact

Anonymity by those undertaking surrogacy is according to the BMA (1996) neither practical nor desired. Personal contact before and after the surrogacy procedures is now advocated. Unfortunately evidence for these views is controversial. According to Pannor and Baran (1984) psychological damage is possible in 'open' adoption cases, where the mother participates in the placement process. Whereas Waddoups (1991) argues that 'open' adoption mothers tend to be more attached to their pregnancies and were able to come to terms with the loss of their baby better than mothers involved in 'closed' adoption. Fear of not bonding with the adopted child was a predictor of resistance to open adoption in Waddoups' (1994) study. In surrogacy, either could be expected. It is possible that active involvement of both parties in this practice is good for the surrogate, but might lead to conflict of interest for intended parents. For example, a surrogate's decision not to breastfeed and decisions about the delivery may not match those of the intended parents (see also Chaps. 4, 5 and 6 in Part II).

Research on attitudes towards pregnancy, the foetus and baby has shown this can vary enormously between pregnant women (Marteau 1989). Birth mothers offering their baby for adoption are known to suffer grief reactions (Stiffler 1991) even if these adoptions were 'open'. To date questions about adjustment to loss in surrogates and motherhood in intended parents have not been addressed. Future investigations are expected to have a major impact on the practice of reproductive technologies in general (including people's long-term feelings about giving up gametes or embryos to others) and surrogacy in particular. To date few systematic in-depth studies have addressed the procedures and long-term impact of surrogacy on the parents and extended families.

Surrogate Motherhood Arrangements

When individuals embark upon a surrogate motherhood arrangement, they need to set a number of objectives because it could potentially be a hugely complex process. A number of factors need to be considered, including a foetal abnormality detected in the pregnancy, a miscarriage or medical abortion, the birth of a disabled infant, stillbirth and a neonatal death. Similarly, the surrogate mother may end up seriously compromised during the pregnancy, psychologically or medically and she could die as a result of the treatment or the birth of the baby. Lastly, one or both of the intended parents could die after the agreement was set up, but before a PO is granted. Although in the United Kingdom such agreements are often mediated by lawyers, they are legally not binding. Despite this it is usually advised that an agreement is set up. If nothing else, it prepares both parties for the possibility that things can go wrong (as shown in Chap. 4).

Considering the increase in gestational surrogacy over the last few decades, it is somewhat surprising that surrogacy is not routinely mentioned in obstetrics textbooks, and clinical guidelines specific to surrogate arrangements are rare (Burrell and Edozien 2014). Burrell and Edozien (2014) described the existing clinical guidelines as including the International Federation of Gynecology and Obstetrics (IFCOG 2008) American College of Obstetricians and Gynecologists (ACOG 2008), European Society of Human Reproduction (ESHRE 2005; see Shenfield et al. 2005), ASRM, (ASRM 2013) and the HFEA (HFEA 2013). The next chapter describes the organisations involved in surrogacy in more depth.

Summary

Surrogate motherhood is generally well defined. It fulfils a shifting population need and is now no longer exclusively an option for the traditional medically infertile heterosexual couple, but has become a method of family building for individuals who are in new relationships when they are of a reproductively incompetent age; same-sex relationships; or for single older women, all of whom may have been reproductively capable at some point. It is one of the more unusual forms of human reproduction. The focus in surrogate motherhood to overcome involuntary childlessness for this myriad of potential commissioning couples and individuals is primarily on their needs, less so the needs of the surrogate mother and not at all that of the child(ren) resulting from these arrangements. The immediate family experiencing the loss and gain of the surrogate baby, as well as the extended family of grandparents, aunts and uncles, and the wider community may or may not welcome the addition (in commissioning) or the loss (in surrogate) of the surrogate-born person to the network. These issues appear to be secondary or assumed to be problem free. The next chapter will discuss the wider consequences of surrogate motherhood by considering theoretical perspectives and population attitudes to this fragmented family building practice.

References

Addelson, K. (1990). Some moral issues in public problems of reproduction. *Social Problems, 37*, 1–17.

American College of Obstetricians and Gynecologists. (2008). Surrogate motherhood. ACOG committee opinion no. 397. *Obstetrics and Gynaecology, 111*, 465–470.

Appleton, T. (2001). Surrogacy. *Current Obstetrics & Gynaecology, 11*, 256–257.

ASRM. (2013). Ethics Committee of the American Society of Reproductive Medicine. Consideration of the gestational carrier: A committee opinion. *Fertility & Sterility, 99*, 1838–1841.

ASRM. https://www.asrm.org/Templates/SearchResults.aspx?q=surrogacy&p=1. Accessed 18 Jan 2016.

Bartholet, E. (1993). *Family bonds: Adoption and the politics of parenting.* New York: Houghton.

Bartholet, E., Draper, E., Resnic, J., & Geller, G. (1994). Rethinking the choice to have children. *The American Behavioral Scientist, 37*(8), 1058–1073.

Batson, C. (2014). Altruism or egoism. In *The altruism question toward a social-psychological answer* (pp. 6–7). Hoboken: Taylor and Francis.

Beeson, D., Darnovsky, M., & Lippman, A. (2014). *What's in a name? Variations in terminology of third party reproduction.* Paper presented at the International

Forum on Intercountry Adoption and Global Surrogacy, The Hague: ISS (11–13 August) cited in: Cheney, K. (2016). Preventing exploitation, promoting equity: Findings from the International Forum on Intercountry Adoption and Global Surrogacy 2014. *Adoption & Fostering, 40*(1), 6–19.

Bello, F., Akinajo, O., & Olayemi, O. (2014). In-vitro fertilization, gamete donation and surrogacy: Perceptions of women attending an infertility clinic in Ibadan, Nigeria. *African Journal of Reproductive Health, 18*(2), 127–133.

Bernardes, J. (1993). Responsibilities in studying post-modern families. *Journal of Family Issues, 14,* 35–49.

Blackburn-Starza, A. (2016a, September 14). High court refuses single father's application for a parental order. *Bionews,* p. 819.

Blackburn-Starza, A. (2016b, May 23). Single father wins surrogacy human rights ruling. *Bionews,* p. 852.

Blyth, E. (1994). "I wanted to be interesting. I wanted to be able to say 'I've done something with my life'": Interviews with surrogate mothers in Britain. *Journal of Reproductive and Infant Psychology, 12,* 189–198.

Blyth, E. (1995). "Not a primrose path": Commissioning parents' experiences of surrogacy arrangements in Britain. *Journal of Reproductive and Infant Psychology, 13,* 185–196.

Braverman, A., & Corson, S. (2002). A comparison of oocyte donors' and gestational carriers/surrogates' attitudes towards third party reproduction. *Journal of Assisted Reproduction and Genetics, 19*(10), 462–469.

Brazier, M., Campbell, A., & Golombok, S. (1998). *Surrogacy review for health ministers of current arrangements for payments and regulation.* Report of the review team. Cm 4068. London: Department of Health.

Brinsden, P. (2003). Gestational surrogacy. *Human Reproduction Update, 9*(5), 483–491.

British Medical Association. (1990). *Surrogacy: Ethical considerations. Report of the working party on human infertility services.* London: BMA Publications.

British Medical Association. (1996). *Changing conceptions of motherhood. The practice of surrogacy in Britain.* London: BMA Publications.

Burrell, C., & Edozien, L. (2014). Surrogacy in modern obstetric practice. *Seminars in Fetal and Neonatal Medicine, 19,* 272–278.

Carsten, J. (2004). *After kinship.* Cambridge: Cambridge University Press.

Cassidy, T., & Sintrovani, P. (2008). Motives for parenthood, psychosocial factors and health in women undergoing IVF. *Journal of Reproductive and Infant Psychology, 26*(1), 4–17.

Celebrities. (2016). http://www.therichest.com/expensive-lifestyle/lifestyle/18-celebrities-who-used-surrogacy/. Accessed 20 Jan 2016.

Cherry, M., & Peppin, J. (2005). Part I: North America, United States perspective on assisted reproductive technologies. In *Annals of bioethics: Regional perspectives in bioethics* (p. 34). Lisse: Taylor & Francis e-Library.

Ciccarelli, J. C. (1997). *The surrogate mother: A post birth follow up study.* Unpublished doctoral dissertation, Los Angeles: California School of Professional Psychology.

Ciccarelli, J. C., & Beckman, L. J. (2005). Navigating rough waters: An overview of psychological aspects of surrogacy. *Journal of Social Sciences, 61*(1), 21–43.

Cotton, K. (1985). *Baby Cotton: For love or money.* London: Dorling Kindersley.

Crawshaw, M., Blyth, E., & van den Akker, O. B. A. (2013). The changing profile of surrogacy in the UK – Implications for national and international policy and practice. *Journal of Social Welfare & Family Law, 2,* 1–11.

Darnovsky, M., & Beeson, D. (2014). *Global surrogacy practices.* ISS working paper series. The Hague: International Institute of Social Studies of Erasmus University, p. 54. Hdl.handle.net/1765/77402

Dhont, N., Luchters, S., & Ombelet, W. (2010). Gender differences and factors associated with treatment-seeking behaviour for infertility in Rwanda. *Human Reproduction, 25,* 2024–2030.

Dhont, N., van de Wijgert, J., Coene, G., Gasarabwe, A., & Temmerman, M. (2011). 'Mama and papa nothing': Living with infertility among an urban population in Kigali, Rwanda. *Human Reproduction, 26*(3), 623–629.

Dimka, R. A., & Dein, S. L. (2013). The work of a woman is to give birth to children: Cultural constructions of infertility in nigeria. *African Journal of Reproductive Health, 17*(2), 102–117.

Dyer, S. J., Abrahams, N., Mokoena, N. E., & Van der Spuy, Z. M. (2004). 'You are a man because you have children': Experiences, reproductive health knowledge and treatment-seeking behaviour among men suffering from couple infertility in South Africa. *Human Reproduction, 19*(4), 960–967.

Dyer, S. J., Abrahams, N., Mokoena, N. E., Lombard, C. J., & van der Spuy, Z. M. (2005). Psychological distress among women suffering from couple infertility in South Africa: A quantitative assessment. *Human Reproduction, 20*(7), 1938–1943.

Dyer, S., Mokoena, N., Maritz, J., & van der Spuy, Z. (2008). Motives for parenthood among couples attending a level 3 infertility clinic in the public health sector in South Africa. *Human Reproduction, 23*(2), 352–357.

Gaither. (2015). *Using a surrogate mother: What you need to know.* WebMD. Retrieved April 6, 2014.

Golombok, S. (1992). Psychological functioning in infertility patients. *Human Reproduction, 7*(2), 208–217.

Hanafin, H. (1984). *The surrogate mother: An exploratory study*. Ph.D. Dissertation, California School of Professional Psychology, Los Angeles, CA. *Dissertation Abstracts International, 45*(10-B), 3335–3336.

Hanafin, H. (1999). Surrogacy and gestational carrier participants. In L. Hammer-Burns & S. C. Covington (Eds.), *Infertility counselling* (pp. 375–0388). Pearl River: Parthenon.

Harrison, M. (1990). Psychological ramifications of surrogate motherhood. In N. L. Stotland (Ed.), *Psychiatric aspects if reproductive technology*. Washington, DC: American Psychiatric Press, Inc.

Heinicke, C. M. (1995). Determinants of the transition to parenting. In M. Bornstein (Ed.), *Handbook of parenting, Status and social conditions of parenting* (Vol. 3). Mahwah, NJ: Erlbaum.

Henaghan, M. (2013). International surrogacy trends: How family law is coping. *Australian Journal of Adoption, 7*(3), 1–24.

HFEA. (2008). Human Fertilisation and Embryology Act. Legislation.gov.uk/ ukpga/2008/22/contents. Accessed 25 May 2016.

HFEA. (2009). S1, 1.3. https://www.gov.uk/government/uploads/system/ uploads/attachment_data/file/268039/surrogacy.pdf. Accessed 23 May 2016.

HFEA. (2013). *Code of practice* (8th ed.). London: HFEA.

HFEA. http://www.hfea.gov.uk/fertility-treatment-options-surrogacy.html. Accessed 17 Jan 2016.

Hohman, H. H., & Hagan, C. B. (2001). Satisfaction with surrogate mothering: A relational model. *Journal of Human Behaviour in the Social Environment, 4*, 61–84.

Hornsey, K., Smith, N., Norcross, S., Ghevaert, L., & Jones, S. (2015). *Surrogacy in the UK: 'Myth busting and reform'*. Report of the Surrogacy UK Working Group on Surrogacy Law Reform, Surrogacy UK, November 2015.

Human Fertilisation and Embryology Act. (1990). London: Her Majesty's Stationary Office.

Human Fertilisation and Embryology Act. (2016). *Egg and sperm donation in the UK: 2012–2013*. http://www.hfea.gov.uk/docs/Egg_and_sperm_dona-tion_in_the_UK_2012-2013.pdf. Accessed 2 May 2016.

Humphrey, M. (1977). Sex differences in attitude to parenthood. *Human Relations, 30*, 737–749.

Inhorn, M. (2003). Global infertility and the globalization of new reproductive technologies: Illustrations from Egypt. *Social Science & Medicine, 56*(9), 1837–1851.

International Federation of Gynecology and Obstetrics & FIGO. (2008). Committee for the Ethical Aspects of Human Reproduction and Women's

Health. Surrogacy. *International Journal of Gynecology & Obstetrics, 102,* 312–313.

Jaggar, A. M. (1994). *Living with contradictions: Controversies in feminist social values.* Boulder: Westview.

Kleinpeter, C. B. (2002). Surrogacy: The parents' story. *Psychological Reports, 91,* 135–145.

Krishnan, V. (1994). Attitudes toward surrogate motherhood in Canada. *Health Care for Women International, 15*(4), 357.

Langridge, D., Connolly, K. J., & Sheeran, P. (2000). Reasons for wanting a child. *Journal of Reproductive and Infant Psychology, 18*(4), 321–338.

Lasker, J. N., & Borg, S. (1987). *In search of parenthood: Coping with infertility and high-tech conception* (pp. 11–11). Boston: Beacon.

Levine, N. (2008). Alternative kinship, marriage and reproduction. *Annual Review of Anthropology, 37,* 375–389.

Marteau, T. (1989). The psychological costs of screening. *British Medical Journal, 299*(6698), 527.

Matetakufa, S. N. (1998). Our own gift. *New Internationalist Magazine, 303,* 11.

Miall, C. (1989). Reproductive technology versus the stigma of involuntary childlessness. *Social Care Work, 70*(1), 43–50.

Millbank, J., Stuhmcke, A., & Karpin, L. (2016). Embryo donation and understanding of kinship: The impact of law and policy. *Human Reproduction.* doi:10.1093/humrep/dew297.

Ombelet, W., Cooke, I., Dyer, S., Serour, G., & Devroey, P. (2008). Infertility and the provision of infertility medical services in developing countries. *Human Reproduction Update, 14*(6), 605–621.

OTA. (1988). *Office of Technology Assessment. Infertility: Medical and social choices.* Washington, DC: Government Printing Office.

Pannor, R., & Baran, A. (1984). Open adoption as standard practice. *Child Welfare, 63,* 245–250.

Petitfils, C., Sastre, M., Sorum, C. P., & Mullet, E. (2017). Mapping people's views regarding the acceptability of surrogate motherhood. *Journal of Reproductive and Infant Psychology, 35*(1), 65–76.

Poote, A., & Van den Akker, O. (2009). British women's attitudes to surrogacy. *Human Reproduction, 24*(1), 139–145.

Postgate, J. N. (1992). *Early Mesopotamia society and economy at the dawn of history* (p. 105). Routledge. isbn:0-415-11032-7.

Purewal, S., & van den Akker, O. B. A. (2007). The sociocultural and biological meaning of parenthood. *Journal of Psychosomatic Obstetrics and Gynaecology, 28*(3), 79–86.

Ragone, H. (1994). *Surrogate motherhood: Conception in the heart.* Boulder: Westview Press.

Schenker, J. (1997). Infertility evaluation and treatment according to Jewish law. *European Journal of Obstetrics, Gynecology and Reproductive Biology, 71,* 113–121.

Schmulker, I., & Algen, B. (1989). The terror of surrogate motherhood: Fantasies, realities and viable legislation. In J. Offerman-Zuckerberg (Ed.), *Gender in transition: A new frontier.* New York: Plenum.

Sharma, B. (2006). Forensic considerations of surrogacy – An overview. *Journal of Clinical Forensic Medicine, 13,* 80–85.

Shenfield, F., Pennings, G., Cohen, J., Devroey, P., de Wert, G., & Tarlatzis, B. (2005). ESHRE task force on ethics and law 10: Surrogacy. *Human Reproduction, 20*(10), 2705–2707.

Soundby, J. (1997). Infertility in the Gambia: Traditional and modern health care. *Patient Education and Counselling, 31*(1), 29–37.

Stiffler, L. H. (1991). Adoption's impact on birthmothers: "Can a mother forget her child?". *Journal of Psychology and Christianity, 10,* 249–259.

Strathern, M. (2002). Still giving nature a helping hand? Surrogacy: A debate about technology and society. *Journal of Molecular Biology, 319,* 985–999.

Surrogacy Arrangements Act. (1985). London: Her Majesty's Stationary Office.

Tangri, S., & Kahn, J. (1993). Ethical issues in the new reproductive technologies: Perspectives from feminism and the psychology profession. *Professional Psychology: Research and Practice, 24,* 271–280.

Thompson, C. (2005). *Making parents: The ontological choreography of reproductive technologies.* Cambridge, MA: MIT Press.

Thorpe, K., Dragonas, T., & Golding, J. (1992). The effects of psychosocial factors on the mother's emotional well-being during early parenthood: A cross-cultural study of Britain and Greece. *Journal of Reproductive and Infant Psychology, 10*(4), 205–217.

Utian, W., Sgheenan, L., Goldfarb, J., & Kiwi, R. (1985). Successful pregnancy after in-vitro-fertilization-embryo transfer from an infertile woman to a surrogate. *New England Journal of Medicine, 313,* 1351–1352.

van Balen, F., & Trimbos-Kemper, T. C. (1995). Involuntarily childless couples: Their desire to have children and their motives. *Journal of Psychosomatic Obstetrics and Gynecology, 16*(3), 137–144.

van den Akker, O. B. A. (1998). Functions and responsibilities of organizations dealing with surrogate motherhood in the UK. *Human Fertility, 1,* 10–13.

van den Akker, O. B. A. (1999). Organizational selection and assessment of women entering a surrogacy agreement in the UK. *Human Reproduction, 14*(1), 101–105.

van den Akker, O. (2000). The importance of a genetic link in mothers commissioning a surrogate baby in the UK. *Human Reproduction, 15*(8), 1849–1855.

van den Akker, O. (2001). The acceptable face of parenthood: The relative status of biological and cultural interpretations of offspring in infertility treatment. *Psychology, Evolution and Gender, 3,* 137–153.

van den Akker, O. (2002). *The complete guide to infertility. Diagnosis, treatment, options.* London: Free Association Press.

van den Akker, O. B. A. (2003). Genetic and gestational surrogate mothers' experience of surrogacy. *Journal of Reproductive and Infant Psychology, 21*(2), 145–161.

van den Akker, O. B. A. (2005). A longitudinal pre pregnancy to post-delivery comparison of genetic and gestational surrogate and intended mothers: Confidence and genealogy. *Journal of Psychosomatic Obstetrics and Gynecology, 26*(4), 277–284.

van den Akker, O. (2007). Psychosocial aspects of surrogate motherhood. *Human Reproduction Update, 13*(1), 53–62.

van den Akker, O. (2012). *Reproductive health psychology* (pp. 162–165). Chichester: Wiley.

van den Akker, O. B. A. (2016). Reproductive health matters. *The Psychologist, 29*(1), 2–4.

van den Akker, O. B. A., Camara, I., & Hunt, B. (2016a). Together … for only a moment' British media construction of altruistic non-commercial surrogate motherhood. *Journal of Reproductive and Infant Psychology, 34*(3), 271–281.

van den Akker, O. B. A., Fronek, P., Blyth, E., & Frith, L. (2016b). 'This neo-natal ménage à trois' British media framing of transnational surrogacy. *Journal of Reproductive and Infant Psychology, 34*(1), 15–27.

Waddoups, S. (1991). *Resistance to open adoption.* Unpublished manuscript, Department of Social Work, University of Nevada, Las Vegas.

Waddoups, J. (1994). Open adoption, human capital formation, and uncertainty. *Journal of Family and Economic Issues, 15*(1), 5–21.

Warnock, M. (1984). *A question of life: The Warnock report on human fertility and embryology.* Oxford: Blackwell.

Women & Health, Health Information and More. (2016). http://www.womens-health.co.uk/surrogacy-the-commissioning-parents.html. Accessed 23 May 2016.

2

Theoretical Perspectives and the Social Context

Surrogate motherhood conflicts with traditional kinship ideology and has therefore been approached with caution, concern and disdain. This chapter considers the challenges posed by all parties involved in surrogate motherhood triads through the lens of a number of selected theoretical frameworks, addressing how individuals reconcile their unusual reproduction with the traditional family. The theories, media influences and surveys of population attitudes which contextualise surrogacy differently demonstrate that the focus tends to favour tradition rather than modern family building demanded by shifting population needs. Modern family building needs are propagated but not adequately supported by governments. More effective future theoretically based research could frame the needs of those involved in unusual reproduction better.

Setting the scene for a theoretical model for surrogate motherhood practices is complicated because there is nothing "natural" about it (sex is not involved), as was expressly exemplified by the early terminology used; artificial (made or produced by humans, rather than naturally occurring) reproduction. Donor insemination and in vitro fertilisation introduce a "difference" into all assisted conception families which differentiates them from traditional families conceiving within a private unit. Despite these differences, families using third-party-assisted conception tend to

© The Author(s) 2017
Olga B.A. van den Akker, *Surrogate Motherhood Families*,
DOI 10.1007/978-3-319-60453-4_2

go through great lengths to claim a connection with the conceptus that is as similar to traditional conception as possible, even though the process with third-party involvement makes these conceptions radically different from the traditional family. Surrogate motherhood, in addition to donor (or commissioning father's sperm) insemination and in vitro fertilisation, adds yet another layer of "difference" within these triads and in surrogate motherhood cases, it tends to be the surrogates, not the commissioning parents, who are questioned about their participation in this difference as they are the birthing mothers of babies nurtured biologically by them. Intended commissioning parents cling to aspects of the arrangement that emphasise the importance of genetics if they can – when they have used either the commissioning mothers' or the fathers' gametes or both. Where this is not possible, the importance of all biological aspects is demeaned and marginalised, in favour of the social parenting roles and responsibilities. The following theories incorporate surrogacy differently and try to explain the importance of the changing social contexts in which surrogate and commissioning intended parents reconcile their differences from the traditional family.

Theories

Traditional Family Theory

Traditional family theory broadly grew out of utilitarian perspectives, drawing on psychological, sociological, philosophical and historical contexts (van den Akker 2001). A traditional family consists of a mother, father and one or more children conceived within the institution of marriage. According to Goode (1964) early philosophers such as Confucius believed happiness and prosperity would come to people if they followed filial obligations within the family. Certainty of paternity and maternity separates civilised society from uncivilised society, reinforcing the importance of biological ties, genetic certainty and the sanctity of marriage, which keep these certainties together. These historical views of the family helped shape further the social debates of the functions of the family by Engels (1984[1972]) and led to later structuralism-functionalist family theory which emphasised the social uses of family units—kept together through biological (genetic) relatedness. Biological facts and genetic relationships were therefore important parameters of kinship and the family (Morgan 1971), and biological definitions of pathology through medicine as a form of social control have been documented by Foucault (1973, 1976). Infertility and its treatment focused on biological (medical) solutions to these problems, and these more modern medical discourses of kinships, led to new cultural definitions of sexuality, reproduction, parenthood and the family, although these discourses are Western and not universal facts (Malinowsky 1923).

By the 1960s and 1970s a renewal of approaches to family theory was necessary to incorporate new fields of study (e.g. marital violence) and the increasing occurrence of sex outside of marriage. Increasing rates of divorce, the independence of women in the workplace, decreases in marriages and pregnancies, increases in co-habitation, single parenthood and gay and lesbian parenting relationships all decreased the predominance of the traditional family. The traditional two heterosexual parents with children all living together in harmony under one roof was therefore challenged, and now, decades later, only fleetingly reflects the norm. Despite these social changes, surrogate motherhood does not sit well within modern family theory, perhaps because the fragmentation of procreation and the relinquishment of a baby threaten social stability too much for comfort.

Political Feminist Models

Feminists generally accept that autonomy over women's reproductive decisions should lie in their own hands. This means that choosing whether to have children, when to have these and what to prioritise in life is their decision. This belated equality with men is reinforced in Western cultures with policies encouraging independence in lifestyle choices to further the educational and economic aspirations of half the population. However, these same policies do not facilitate the consequences of abortion, contraceptive use and leaving childbearing until it is too late. Too little state assisted help is available to women to have children after they focused on education, careers and financial security. Similar politically driven emancipation measures were evident in the former Soviet Union, with abortion first legalised in Russia in the 1920s (then withdrawn during Stalin's rule in 1936 and reinstated by Khrushchev in 1955) to encourage women to enter the workforce, but no efforts were made to supply contraceptive services (Field 2007). Encouraging freedom but then taking away the choice to (or when to) parent at a later stage is poor practice. The same poorly considered principles apply to single women, single men and same sex couples who are given a legitimate place in some societies but are politically not facilitated to build families as state funding (e.g. NHS supporting the United Kingdom) is considerably less for same sex and single people seeking to build a family (HFEA 2014, p. 20).

Feminist Theories

Surrogate motherhood is challenging to feminists' ways of thinking too. From the commissioning perspective, Andrews (1988) notes that feminism paved a way to third-party-assisted conception using donated gametes/embryos or surrogacy, because women who exercised their rights to be educated and develop themselves into the workforce were left with reduced childbearing capital. Hence third-party-assisted conception provides a way around the inevitable shortfall in reproductive capacity. Instigating or commissioning a surrogate mother is therefore one of several solutions for feminists exercising their rights. From the surrogate mother's perspective, surrogate mothers are also doing exactly what feminism stands for—exercising control over their reproductive bodies. The fact that a surrogate does not become a social mother is not unlike the choice many fathers make to care or not to care for the child. It is a choice, just like having an abortion or remaining child-free would be a choice. These same feminist principles contain provisos that surrogates should be able to refuse selective feticide, abortions and amniocentesis during pregnancy regardless of what the contracts and legislation stipulate, because she should control her own body. Theoretically, surrogacy therefore allows a woman to choose to do what she wants with her body. Although feminists argue that any woman should be in control of her own body and no one has a right to deny them that, these basic human rights to have reproductive choice need to be weighed against the potential harms that could come to themselves, the children and society from these rights. Any reproductive rights should not be "overridden by symbolic harms or speculative risks to potential children" (Andrews 1988, p. 72).

On the other hand, feminists are concerned that surrogate mothers could be exploited as breeding stock and thereby used and demeaned. This feminist position on surrogacy is firmly set against the practice calling for legal bans. Andrews (1988) lists three main principles underlying this position: (1) the symbolic harm to society in commercial surrogacy; (2) the potential risks to the surrogate mothers and finally (3) the potential risks of commercial surrogacy to the surrogate child, as shown in Fig. 2.1.

Regarding the principle of the symbolic harm to society in commercial surrogacy, the objection is one of buying and selling children. This is a fundamental principle which has its roots in many other approaches to surrogacy as evidenced by, for example, British legislation that bans commercial surrogacy. However, British legislation allows altruistic surrogacy which is subject to some ambiguity regarding "expenses" paid for the inconvenience or the "job." The feminist objection also does not sit well alongside the right to abortion which feminists have fought for, for so long, which in effect has been emotively described by those against it as the killing of children. This symbolic harm is therefore applied in

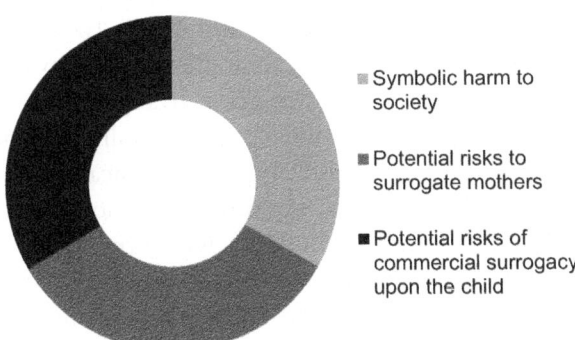

Fig. 2.1 Principles underlying feminist theories opposing surrogate motherhood practices

surrogacy but not abortion. Symbolic harm positions are also riddled with emotive language and rely on tradition in a non-traditional, non-normative context. It is not well supported theoretically or empirically.

The second principle of the potential physical and psychological risks to the surrogate mothers is one that is more difficult to consolidate because there are risks associated with all other aspects of women's reproductive choices, including abortions and caesarean sections (C-sections). Making the wrong decision in retrospect is an inherent risk in having a choice. As the chapter on surrogate mothers shows (Chap. 4), surrogates do not change their minds very often and tend to cognitively restructure their thinking about pregnancy and motherhood over the course of the arrangement, which, if successful, then places them in a relatively favourable position to relinquish the baby. This cognitive processing requires help and direction from others and is known to be encouraged by surrogate agencies in the United States and the United Kingdom respectively (Ragone 1994; van den Akker 2001). It is not in the interests of women and their reproductive freedom to strive for government control over what they can and cannot do with their bodies. As long as the arrangement was entered willingly and fully informed, the woman should keep control. However, we are now aware of the many cases of potential exploitation of ill-informed illiterate women in developing countries entering into cross-border surrogacy arrangements (see Chap. 8). With the more recent trend of what has been termed baby farming, some governments have initiated bans on cross-border commercial surrogacy (for more details on this, see Chap. 9 on ethical and moral, and Chap. 10 on legal issues in surrogacy). There are therefore contextual differences between the developed and developing world and surrogacy arrangements. Although in both cases there is undisputed evidence that overall the commissioning couples are substantially better off and better educated than surrogate mothers, in the developing world exploitation is more overtly apparent which is not generally found in the West, where the women themselves approach the agents to offer their services as surrogates. For example, in India, Thailand or Cambodia on the other hand, scouts seek out poor communities to recruit deprived women to their clinics. In the first case it is therefore potentially patronising to talk about informed consent, whereas in the latter case it may be patronising not to do so.

There is also evidence of another kind of control. Rivkin-Fish (2013) mentions limits to autonomy assumed by surrogate mothers. Commissioning mothers in Russia tend to control every aspect of the surrogate mother ranging from conception to delivery and every stage in between. Many keep the surrogates within their own homes or in nearby apartments dedicated to the surrogate in order to monitor and control them. The women are never equal since there is an economic disparity between the surrogate and commissioning women, which may well present pressures not liberties to, for example, leave their own children for the duration of the arrangement in order to get the promised money. Here it can still be seen as objectification of women, even if the surrogates have ultimate control with the law leaving them in charge of the fate of the child (see Chap. 8). Surrogate mothers are therefore in effect not autonomous but vulnerable to exploitation (Berkhout 2008; Ryan 2009) although they themselves can also exploit the commissioning parent(s). More theoretically driven research is needed to unravel the true extent of autonomy and exploitation between the parties in the arrangement (see also the section on geographical theories), and cross-national comparisons need to be considered independently as views, cultural values, economic state support and religion will influence them differently. International surrogate and commissioning parents' advocacy should focus on respect. The role of men also needs to be considered as it may be that their roles as providers and protectors and fathers are becoming less relevant in modern society, but their desires to build families remain strong.

Finally, the potential risks of commercial surrogacy to the surrogate child principle is probably the most important. Feminists are concerned about the welfare of children, and in surrogacy, the position is that mothers are best placed to make decisions for their children, not governments or legislation. The most profound aspect of this principal objection to surrogacy concerns the notion that a child is bought and sold, it has become a commodification. When a genetic link binds the commissioning parents to the baby, there is less concern about commodification—it is deemed a sufficient indicator of parental "merit" (Andrews 1988, p. 77). When there is no such link, as in genetic surrogacy using donor sperm or surrogacy using donated embryos, commissioning parents choose the gametes. Here the commissioning parent(s) opt in and opt out of several characteristics in a future child which commodifies the process (see Chap. 3). It is also argued that the surrogate baby is not under any psychological stress as it is placed with commissioning parents immediately or soon after birth. Once with the commissioning parent(s), it is not sold, but cared for within the new family unit. Nevertheless, parents' expectations of these (expensive) children may be too much for the resultant children—who become adults—to live up to. Fortunately, there is no evidence that IVF parents, for example, who also tend to spend huge amounts of money on fertility treatments, expect more of their child(ren) than traditional parents. There is also no evidence that surrogates' own children feel threatened by thinking they too could be relinquished to another couple. It would undermine the surrogate mothers' ability to mother effectively if these reasons were used as potential risks of commercial surrogate motherhood, in a similar way that an abortion could affect existing children in a family. They too might think that their mother might "send them to heaven" just like the foetus aborted for social or medical reasons (Fletcher 1982). An aborted conceptus for medical reasons was in fact intended to be part of the family and is now dead—not even living with other people (Andrews 1988, p. 77), potentially resulting in grief reactions within the family, but a surrogate baby was never meant to join the surrogate mother's family.

Geographical Theories

Geographical theories or positions reflect the interactions and effects between people and the physical environment. It concerns the role of the environment and how it determines or influences behaviours, which differ according to numerous factors, how we impact upon it and how we "interact" with the environment. Early geographical theories looked to the power of the earth's influences on people (see, for example, Strabo, as described in Jones 1917). Over time, and with increasingly sophisticated science and technology, views have shifted more towards what people did to the environment, which developed into close links between geography and sociology (see, for example, Weber (translated) 2003). Weber (2003) focused, for example, on politics, capitalism and the influence of religion (to control populations and instil order in what could otherwise be chaos) and economics upon the environment. Cultural determinism also borrowed from Goethe and others; and from the Romanticism era (where geography was thought to influence people, and culture worked in harmony with the societies); and anthropology (for example, Steward, 1955's theory of culture change) which discusses the way in which cultures change and are induced by a need to adapt to the changing environment.

This is not dissimilar to more recent attempts to relate the influence of the media upon behaviours. In media theory political arrangements are determined by media images and media input determines economic and political arrangements (Negrine 2005). Media theory drifted into the undeniably subtler approach which saw the best developments in human geography adapting a balance of power standpoint between the environment and its people. A number of issues amongst such theories are relevant to surrogate motherhood, including environmental factors dictating social mobility and events; cultural factors dictating our actions upon the environment, as actors or within the limitations of what the environment has to offer called "possibilism"; and "probablism" or cultural ecology, which sits somewhere between the previous two theoretical positions, accepting that in some situations the influence of the environment determines people's actions and vice versa.

An early geographical perspective developed by feminist geographer Massey (1994) focuses on power (who moves and who does not, Masey 1993); place (by what social process place is constructed, Harvey 1996) and positionality (power relations between commissioning and surrogate mothers, e.g. Mountz 2011), referring to social relations infused with power, meaning and symbolism (Massey 1994). The social geometry is a power-laden geometry where who moves and who stands still is determined by who holds the power with unequal resources at the core of this inequality. Social hierarchies are reinforced by keeping the weak weaker, and restricting social mobility to those with power (Massey 1993, p. 62). This perspective argues that the "flow of capital influences the construction of the geographical configuration of place" (Deomampo 2013). Communication, travel and technology allow for the flowing of reproductive tourism networks, with high-income countries meeting low-income countries to satisfy demands for reproductive labour

(Harvey 1996, p. 293). According to Deomampo (2013), geographical analyses of reproductive tourism implicate space, place and mobility, drawing also on feminist theory. Feminist and geographical theory together can accommodate the contradictions and complexities of (cross-border) surrogacy, including the surrogate and commissioning parents' times of movement and mobility, stillness and immobility and anticipation. In her framework, both parties are exploiters and exploited or agents and victims, which has some foundations in national arrangements too (see, for example, van den Akker 2007; Ragone 1994). According to this model, surrogate and commissioning parenthood are bound by the same rules of "space, mobility and immobility reflecting global, social and economic hierarchies" (Deomampo 2013, p. 521). Cross-border surrogacy magnifies what is apparent in national surrogacy; the agents and agencies involved create spaces and places of production and consumption. The meanings of these are likely to change over time and the relationships are not dichotomous but shifting along a continuum with news media a key player in framing these meanings. Nevertheless, reproductive inequalities remain unbalanced, reflecting the global economic disparities of our time (Deomampo 2013).

The concept of "precarity" or employment deficient in labour security when compared with society's standards for a decent job (Thornley et al. 2010, p. 114) is appropriately applied to surrogate mothers who do surrogacy as "work," i.e. commercial surrogacy. Precarity of work tends to be undertaken by individuals who are poorly educated, deprived and often relatively young (Standing 2014), a profile characteristic of surrogate mothers (van den Akker 2007). Surrogate motherhood is also marked by social stigma in many societies in which women take these commercially arranged surrogate motherhood jobs. The precariat (individuals defined by this work) have no central place or union to support and protect their welfare (Thornley et al. 2010, p. 30), and are disposable employees detached from the job they perform (Walkerdine and Jimenez 2012). Precarity is also casual, devoid of security, job identity, predictability, social benefits such as health, safety, pensions or offering psychological welfare. In surrogate motherhood, this alienation from the surrogate arrangement/job may lead to further alienation of their identity as is reported in other precariat (see, for example, Standing 2014), with further negative effects upon their social status in society and their psychological health (Bluestone and Harrison 1982). This is important because precarity goes against the principles of progress of the social and welfare state and the ideology of democracy (Thornley et al. 2010, p. 23). In surrogate motherhood arrangements, the long-term medical health of the surrogate mother may also be compromised which goes against all ethical principles to do no harm.

The Social Context

The theories presented so far provide a much needed backdrop to the research into surrogacy. The most critical aspect of theoretical principles and frameworks is that the context is considered. Exporting applications of theories across different traditions is not always appropriate or effective (Funk 2004). However, re-evaluating nationally important concepts affecting reproduction and reproductive rights and seeking to apply these cross-culturally is a mammoth task that needs to be tackled. Critics of past research into surrogate motherhood (for example, Teman 2008) argue that cultural settings should not come into the scientific enquiry. She argues, for example, in a review article that much psychosocial research (not ethnographic research) is uncritical and subject to "representing a cultural text on the norms and values of Western culture and reveals how Western cultural assumptions impact scientific research" (Teman 2008, p. 1105). Teman goes further and assumes that psychosocial research using scientific methods "try to authoritatively prove that women who willingly relinquish a child in surrogacy are not 'normal.'" This is a naïve assumption. Scientific research makes no authoritative assumptions; it explores the possibilities of how and why so many surrogates successfully enter these agreements and what makes them able to complete the process without much hardship, which is acknowledged as different from adoption hardship, as there is often little choice involved in the latter case. Moreover, the critique is selective as Teman does not cite the same authors she criticises for compartmentalising surrogates in normalcy terms, that most surrogates themselves say they are different, that it requires a "special kind of person" to do this (Ragone 1994; Blyth 1994; van den Akker 2007). Finally, culture-bound beliefs provide the context in which surrogacy takes place, making it meaningless to explore surrogacy outside of its context. According to Munoz Sastre et al. (2016), the acceptability of a behaviour or decision tends to be "situational" rather than all or none. Specific cultural values also reflect the specific legal positions adapted between different countries, with those which rely heavily on, say, the Roman Catholic tradition, legally banning all forms of surrogacy whereas other jurisdictions bestow motherhood to birth, or to genetic mothers or fathers. Laws therefore are intractably related to differing cultural values and any research or theory on surrogate motherhood cannot usefully be developed out of context.

Perceived Acceptability of Surrogacy

Since research into surrogate motherhood is not as prolific as one might have expected because of its controversialist and relatively novel nature embedded within the world of the "new reproductive technologies," it is surprising not more research is commissioned in this area. Surrogate motherhood, alongside all other new reproductive technologies such as gamete and embryo donation, mitochondrial donation, prenatal genetic diagnosis and so on, has the potential to change the footprint of humanity. All are also capable of changing epigenetics, improving the human gene pool and eradicating undesirable traits. However, with such ambitious futuristic possibilities come as many controversies (see, for example, van den Akker 2013, 2016). The relative "hotness" of the topic no doubt assists in the fuelled debates raging in different research papers in different countries. Countries strive for modernity and improving health care contextualised within old and ancient cultural and religious parameters which are paramount to the place surrogate motherhood practices have within these differing cultural settings.

Attitudes Towards Surrogacy

Previous reports on attitudes towards surrogacy (Appleton 2001; Banerjee and Basu 2009; Brinsden et al. 2000; Chilaoutakis et al. 2002; Dutney 2007; Genius et al. 1993; Schenker 2005; Stuhmcke 1996) have indicated a number of factors which may influence attitudes and beliefs about surrogacy. These factors include traditional definitions of parenthood, religious beliefs/restrictions, media portrayal, governmental laws regulating commercial or non-commercial surrogacy, medical conditions and perceived behavioural control. Such attitudes and beliefs, in turn, influence their intentions. A recent study of French laypersons reported higher rated moral problems with surrogacy if the surrogate mother was not autonomous, if sperm rather than embryos and if genetic surrogacy was used. Parous individuals and individuals who attended church regularly rated more moral problems than childfree non-church attenders, but the difference in acceptability between genetic and gestational surrogacy was small (Petitfils et al. 2017). Surprisingly, altruistic surrogacy using a known surrogate was reported as more morally problematic than commercial surrogacy.

Few theoretical models have been tested within social sciences research into infertility, treatments to overcome infertility and successful and unsuccessful outcomes (van den Akker 2012). In one study, Poote and van den Akker (2009) tested components of the theory of planned behaviour to understand pertinent factors influencing intentions to become a surrogate. According to the theory of planned behaviour, the intention to perform a particular behaviour is influenced by a number of independent factors, including attitudes towards that behaviour, normative beliefs and motivation to comply with others, and perceived behavioural control (Ajzen 1991). Attitudes towards the behaviours refer to a person's overall evaluation of the behaviour, that is, the beliefs about the positive or negative consequences of performing that behaviour (Ajzen 1991; Armitage and Conner 2001). The normative beliefs and motivation to comply refer to the perceived social pressures exerted by important people and their beliefs about whether or not the behaviours should be performed (Ajzen 1991). Perceived behavioural control refers to how much control the person has over the behaviour and also to how confident s/he feels about carrying out the behaviour. It is determined by situational factors such as anticipated obstacles or impediments, or internal factors such as past experience, that either facilitate or inhibit the execution of the behaviour (Ajzen 1991).

People's attitudes to surrogacy also depend upon their knowledge of surrogacy (Minai et al. 2006) and ethical beliefs about right and wrong

(Brinsden et al. 2000). Many years ago, Genius et al. (1993) reported that 85% of the public opposed social surrogacy if this was used for convenience. Convenience here referred to women who were not willing to go through the inconvenience of pregnancy; who want to maintain their body shape or who do not want to put their careers on hold. Similarly, questions about exploitation arose; should fertile women bear the burden of procreation for infertile women altruistically or for financial gain? (Banerjee and Basu 2009). People's ethical beliefs about rights and wrongs are likely derived from deeply ingrained moral persuasions about what constitutes a family and relatedness (Stuhmcke 1996) and these century-old traditions clearly state that procreation takes place in permanent, marital, heterosexual relationships excluding any additional "outside" help (Dutney 2007). This was the socially accepted belief of a "nuclear family" that people still hold, and deviation from this norm does not easily gain social acceptance (van den Akker 2007), as is repeatedly reported in the press (van den Akker et al. 2016a, b; Markens 2012). For example, there are reports of surrogate mothers' refusal to enter a surrogacy arrangement with gay men or who refuse to hand over the surrogate baby after finding out the commissioning couple were gay at a later stage (see, for example, Asian Correspondent Staff 2016). Here, an American and Spanish gay couple started a legal battle in Thailand against a surrogate mother, who changed her mind about handing the child to them when she found out the couple was gay.

Media Framing

The popular press often report surrogate motherhood cases in negative frames, focusing on cases which go wrong or those which are wrong (Markens 2012). For example, the infamous Baby M case in America in the 1980s (Scott 2009) which went horribly wrong, the baby Gammy case in Thailand (Michael 2014) which involved a deeply questionable ethical and moral "wrong," and more recently a child sexual abuse case of paedophiles abusing Thai surrogate children (Vince 2014) have tainted the practice significantly. At the same time, entertainment programmes about surrogacy (e.g. 'Friends' and 'Baby Mama') showed a more positive side of the practice, leaving the public with the difficulty of interpreting these opposing scenarios. Recently, two studies of British media framing of national (van den Akker et al. 2016a) and international surrogacy (van den Akker et al. 2016b) showed the British press was quite directional in its reporting on these two types of surrogate motherhood arrangements. In the national surrogacy study, (van den Akker et al. 2016a) titles and content were predominantly loss, high alarm and high vulnerability framed, whereas the content was also gain framed. Their focus was on social and legal aspects differentially between the newspaper types, reflecting differing socio-demographic readerships.

International cases of surrogacy have been reported in the British press since 1984, although they were rare until 1996. A clear difference between this and the national surrogacy media reporting in our 2016a study concerns the commissioning parents. In this 2016b study, unlike the national study, 50% of the reports across newspaper types focused on gay couples and 33% on heterosexual couples. Importantly, 24.59% of newspapers framed the reports as gain framed (mainly in relation to the couples or individuals commissioning the arrangement); 20% as loss (mainly focusing on the surrogates and the surrogate babies) framed; 20% as neutral framed; 19% as high alarm and 16% as high vulnerability framed (van den Akker et al. 2016b), emphasising the competing discourses of gain and opportunity or autonomy; loss and inequality or exploitation (Palattiyil et al. 2010; Rotabi and Bromfield 2012; Markens 2012). Interestingly, medical issues concerning either infertility or the medical risks associated with the IVF process were hardly discussed. Instead, journalists reported on international surrogacy as commodification, stressing the large sums of money exchanged and on the social and legal aspects of cross-border surrogacy.

Some religions also oppose third-party reproduction drawing on the sanctity of the marital relationship and that of the embryo (see Chap. 9). For example, Roman Catholicism opposes the use of artificial insemination (including IUI and IVF). Chilaoutakis et al. (2002)'s attitude survey confirmed that churchgoers were less likely to use donated gametes or surrogacy. Other religions, including Judaism, Islam and the Protestant and Anglican denominations, accept assisted conception when used with gametes from the husband and wife (Schenker 2005). A preference for traditional parenthood is therefore unsurprising, considering people's desire to be accepted for their actions by their communities. Surrogacy challenges accepted and normative practices, hence calling for a culture-free approach to the study of surrogacy is not useful. However, challenges to tradition are accompanied by a shifting societal context. The changes in the last few years in, for example, marital laws in the United Kingdom allowing same sex marriages and the decreasing stigma associated with single parenthood mean traditional beliefs need to make way for beliefs reflecting new realities as evident in new policies and practices, although this necessary shift is clearly still in its infancy (Turner 2001; Riggs and Due 2013).

Summary

In summary, this chapter has tried to build a picture of the meagre theoretical contexts devoted to surrogate motherhood. Against a background of traditional family theory, reflecting current and past family values, other theories which directly or indirectly interpret surrogate motherhood practices which have roots in political, sociological and anthropological disciplines are described briefly. Psychologists too have attempted to model some of the behavioural parameters of surrogate motherhood, although it is clear that these attempts are few and far between. This is surprising since it is highly likely that non-traditional family building will be a very large part of our future, with increasing uncertainties about future partnerships, origins, belonging and identity across the surrogates, the commissioning intended parents and the resultant individuals from these arrangements. Building theoretical frameworks to place these activities at the centre would help future participants in these triads to be studied more effectively, particularly with continuous change directing their different future needs. It could also help in framing the needs of these participating triads, and bringing these framed structures to society at large (perhaps via mass media) to help shape positive attitudes towards increasingly non-traditional family building and to contextualise these changes to its political regulators. No matter what the perspective adopted is concerning the rights and wrongs of surrogate motherhood, it is in everyone's interests to prioritise ensuring discrimination faced by the resultant children is minimalised.

References

Ajzen, I. (1991). The theory of planned behavior. *Organizational Behavior and Human Decision Processes, 50*, 179–211.

Andrews, L. B. (1988). Surrogate motherhood: The challenge for feminists. *Law, Medicine and Health Care, 16*(1–2), 72–80.

Appleton, T. (2001). Surrogacy. *Current Obstetrics and Gynaecology, 11*, 256–257.

Armitage, C., & Conner, M. (2001). Efficacy of the theory of planned behaviour: A meta-analytic review. *British Journal of Social Psychology, 40*, 471–499.

Asian Correspondent Staff. (2016, March 24). https://asiancorrespondent.com/2016/03/gay-couple-begin-legal-battle-in-thailand-against-surrogate-over-custody-of-baby/. Accessed 16 Apr 2016.

Banerjee, S., & Basu, S. (2009). Rent a womb: Surrogate selection, investment incentives and contracting. *Journal of Economic Behavior & Organization, 69*, 260–273.

Berkhout, S. G. (2008). Buns in the oven: Objectification, surrogacy and women's autonomy. *Social Theory and Practice, 34*(1), 95–117.

Bluestone, B., & Harrison, B. (1982). *The deindustrialization of America: Plant closings, community abandonment and the dismantling of basic industry.* New York: Basic Books.

Blyth, E. (1994). "I wanted to be interesting. I wanted to be able to say 'I've done something interesting with my life'": Interviews with surrogate mothers in Britain. *Journal of Reproductive and Infant Psychology, 12*(3), 189–198.

Brinsden, P., Appleton, T., Murray, E., Hussein, M., Akagbosu, F., & Marcus, S. (2000). Treatment by in vitro fertilisation with surrogacy: Experience of one British centre surrogacy should pay. *British Medical Journal, 320*, 924–929.

Chliaoutakis, J., Koukouli, S., & Papadakaki, M. (2002). Using attitudinal indicators to explain the public's intention to have recourse to gamete donation and surrogacy. *Human Reproduction, 17*(11), 2995–3002.

Deomampo, D. (2013). Gendered geographies of reproductive tourism. *Gender and Society, 27*(4), 504–537.

Dutney, A. (2007). Religion, infertility and assisted reproductive technology. *Best Practice & Research in Clinical Obstetrics & Gynaecology, 21*, 169–180.

Engels, F. (1984[1972]). *On the origins of the family, marriage, private property and the state.* New York: International Publishers.

Field, D. A. (2007). *Private life and communist morality in Khrushchev's Russia.* New York: Peter Lang.

Fletcher, J. (1982). *Coping with genetic disorders: A guide for counselling.* San Francisco: Harper & Row.

Foucault, M. (1973). *The order of things: An archeology of the human sciences.* New York: Vintage Books.

Foucault, M. (1976). *The history of sexuality* (Vol. 1). Harmondsworth: Penguin Books.

Funk, N. (2004). Feminist critiques of liberalism: Can they travel east? Their relevance in eastern and central Europe and the former Soviet Union. *Signs, 29*(3), 695–726.

Genius, S., Genius, S., & Chang, W. (1993). Public attitudes in Edmonton toward assisted reproductive technology. *Canadian Medical Association Journal, 150,* 701–708.

Goode, W. J. (1964). *The family.* Englewood: Prentice Hall.

Harvey, D. (1996). *Justice, nature and the geography of difference* (p. 293). Oxford: Blackwell.

HFEA. (2014). http://www.hfea.gov.uk/docs/Egg_and_sperm_donation_in_the_UK_2012-2013.pdf. Accessed 3 May 2016.

Jones, H. L. (Trans.). (1917). *The geography of Strabo* (Vol. 1). London: Heinemann.

Malinowsky, B. (1923). The psychology of sex and the foundations of kinship in primitive societies. *Psyche, 4,* 98–128.

Markens, S. (2012). The global reproductive health market: U.S. media framings and public discourses about transnational surrogacy. *Social Science & Medicine, 74,* 1745–1753.

Massey, D. (1993). Power-geometry and a progressive sense of place. In J. Bird, B. Curtis, T. Outnam, G. Robertson, & L. Tickner (Eds.), *Mapping the futures: Local cultures, global change.* London: Routledge.

Massey, D. (1994). *Space, place and gender.* Cambridge, UK: Polity.

Michael, S. (2014, December 23). From being abandoned in Thailand to living in a new house and enjoying family trips to the beach: Baby Gammy celebrates his first birthday after being the centre of an international surrogacy scandal. *Daily Mail Australia.* http://www.dailymail.co.uk/news/article-2884625/From-abandoned-Thailand-living-new-house-enjoying-family-trips-beach-Baby-Gammy-celebrates-birthday-centre-internatuonal-surrogacy-scandal.html#ixzz3dQyroGUy. Accessed 18 June 2015.

Minai, J., Suzuki, K., Takeda, Y., Hoshi, K., & Yamagata, Z. (2006). There are gender differences in attitudes toward surrogacy when information on this technique is provided. *European Journal of Obstetrics, Gynecology and Reproductive Biology, 132,* 193–199.

Morgan, L. H. (1971). Systems of consanguinity and affinity in the human family. Smithsonian contributions to knowledge. Cited in Franklin, S. (1977).

Embodied progress. A cultural account of assisted conception (p. 22). London: Routledge.

Mountz, A. (2011). Spectres at the port of entry: Understanding state mobilities through an ontology of exclusion. *Mobilities, 6,* 317–334.

Munoz Sastre, M., Sorum, P., & Mullet, E. (2016). The acceptability of assisted reproductive technology among French lay people. *Journal of Reproductive and Infant Psychology, 34*(4), 329–342.

Negrine, R. (2005). *Politics and the mass media in Britain* (2nd ed.). London/New York: Routledge/Taylor & Francis.

Palattiyil, G., Blyth, E., Sidva, D., & Balakrishnan, G. (2010). Globalisation and cross-border reproductive services: Ethical implications of surrogacy in India for social work. *International Social Work, 53,* 686–700.

Petitfils, C., Munoz Sastre, M., Sorum, P., & Mullet, E. (2017). Mapping people's views regarding the acceptability of surrogate motherhood. *Journal of Reproductive and Infant Psychology, 35*(1), 65–76.

Poote, A., & Van den Akker, O. (2009). British women's attitudes to surrogacy. *Human Reproduction, 24*(1), 139–145.

Ragone, H. (1994). *Surrogate motherhood: Conception in the heart.* Boulder: Westview Press.

Riggs, D. A., & Due, C. (2013). Representations of reproductive citizenship and vulnerability in media reports of surrogacy. *Citizenship Studies, 17,* 956–969.

Rivkin-Fish, M. (2013). Conceptualizing feminist strategies to Russian reproductive politics: Abortion, surrogate motherhood, and family support after socialism. *Signs: Journal of Women in Culture and Society, 38*(3), 569–594.

Rotabi, K., & Bromfield, N. (2012). The decline in intercountry adoptions and new practices of global surrogacy: Global exploitation and human rights concerns. *Affilia: Journal of Women and Social Work, 27,* 129–141.

Ryan, M. (2009). The introduction of assisted reproductive technologies in the 'developing world': A test case for evolving methodologies in feminist bioethics. *Signs, 34*(4), 805–825.

Schenker, J. (2005). Assisted reproduction practice: Religious perspectives. *Reproductive Biomedicine Online, 10*(3), 310–319.

Scott, E. S. (2009). Surrogacy and the politics of commodification. *Law and Contemporary Problems, 72*(3), Show Me the Money: Making Markets in Forbidden Exchange (SUMMER 2009), pp. 109–146. Duke University School of Law. URL http://www.jstor.org/stable/40647246

Standing, G. (2014). The precariat. *American Sociological Association, 13*(4), 10–12.

Steward, J. H. (1955). *Theory of culture change: The methodology of multilinear evolution* (p. 244). Urbana: University of Illinois Press. isbn:978-0252002953.

Stuhmcke, A. (1996). For love or money: The legal regulation of surrogate motherhood. *Murdoch University Electronic Journal of Law, 3*(1), 1–34.

Teman, E. (2008). The social construction of surrogacy research: An anthropological critique of the psychosocial scholarship on surrogate motherhood. *Social Science and Medicine, 67,* 1104–1112.

Thornley, C., Jeffreys, S., & Appay, B. (2010). *Gobalization and precarious forms of production and employment.* Cheltenham: Edward Elgar Publishing Ltd.

Turner, B. S. (2001). The erosion of citizenship. *The British Journal of Sociology, 52,* 189–209.

van den Akker, O. B. A. (2001). The acceptable face of parenthood. The relative status of biological and cultural interpretations of offspring in infertility treatment. *Psychology, Evolution and Gender, 3*(2), 137–153.

van den Akker, O. (2007). Psychosocial aspects of surrogate motherhood. *Human Reproduction Update, 13*(1), 53–62.

van den Akker, O. (2012). Chapter 10: Overcoming involuntary childlessness and assisted conception. In *Reproductive health psychology* (pp. 162–165). Chichester: Wiley.

van den Akker, O. B. A. (2013). For your eyes only: Bio-behavioural and psychosocial research objectives. *Human Fertility, 16*(1), 89–93.

van den Akker, O. (2016). Reproductive health matters. *The Psychologist, 29*(1), 2–5.

van den Akker, O., Hunt, D., & Camara, I. (2016a). 'Together…for only a moment' British newspaper construction of altruistic non-commercial surrogate motherhood. *Journal of Reproductive and Infant Psychology, 34*(3), 271–281.

van den Akker, O., Fronek, P., Blyth, E., & Frith, L. (2016b). This neo-natal ménage à trois': British media framing of transnational surrogacy. *Journal of Reproductive and Infant Psychology, 34*(1), 15–27.

Vince, M. L. (2014, December 9). Man sexually abused two children born via Thai surrogate, NSW court hears. *ABC News.* http://www.abc.net.au/news/2014-12-9/man-appears-in-court-thai-surrogacy-child-sex-abuse-charge/594954. Accessed 18 June 2015.

Walkerdine, V., & Jimenez, L. (2012). *Gender, work and community after de-industrialization: A psychosocial approach to affect.* New York: Palgrave Macmillan.

Weber, M. (2003). *The protestant ethic and the spirit of capitalism* (trans: Parsons, T.). Mineola: Dover Publications Inc.

3

The Process of Surrogacy

The acceptability and accessibility of surrogate motherhood to support the involuntary childless whilst being reimbursed to build a family depend upon the cultural and social network, local attitudes as well as the institutional values, endorsements and responsibilities, and these are noted in this chapter. Also covered are issues of equity in access to surrogate motherhood and the social contexts necessary to accept non-traditional families. Human-rights-based multidisciplinary/multiprofessional treatments of the parties involved in—and resulting from—these arrangements are increasingly critical for the child's future welfare. Finally, the importance of counselling is emphasised to ensure the long-term welfare of the triads before, during and after these difficult decisions about surrogacy are made is accounted for.

Across the world surrogate motherhood practices vary considerably. For example, surrogacy in the United Kingdom is very different from arrangements in other modern Western cultures such as the United States or from Eastern countries such as India, which also differ greatly from each other. In states in the United States where surrogacy is legal, companies employ surrogate mothers and prescribe what criteria they need to fulfil, how to be a successful surrogate mother and how much will be paid. A tough screening interview is generally carried out which includes

© The Author(s) 2017
Olga B.A. van den Akker, *Surrogate Motherhood Families*,
DOI 10.1007/978-3-319-60453-4_3

a psychological assessment and medical examination. Much is asked about the potential surrogate's social life, her history and her anticipated future plans. If a surrogate does not meet all the criteria, she may be rejected. Commissioning couples pay for the services and are therefore not screened in the same way as surrogates. In the United Kingdom, little of these screening processes are evident. Anyone can become a potential surrogate, provided she has had at least a child of her own. Criteria, though discussed, are lenient, and no thorough examination is needed. To become a surrogate mother in the United Kingdom is therefore different from becoming a surrogate mother in the United States, and surrogates and commissioning parents choose each other. In developing countries such as India, Thailand and Cambodia surrogates are enticed into programmes and are controlled by commercially operated companies or clinics making their experiences difficult to compare. However, in terms of the process, developing countries operate a commercial surrogate motherhood arrangement which is similar to the American principle, and places the British system into a uniquely different position.

Commissioning and Facilitating Family Building

However, the easier it is to become a surrogate mother, the more difficult this arrangement may be for commissioning parents or for the long-term health and wellbeing of the person conceived via surrogacy. For example, the (genetic) surrogate may be a carrier of a disease or genetic physical or mental health condition or have poor nutritional or lifestyle habits which she may not disclose. Similarly, lenient screening, which is generally applied to commissioning couples across the world, may compromise the health and wellbeing of the surrogate mother and the surrogate baby. A commissioning father or mother (or sperm, egg or embryo donor) may also be a carrier of a disease or genetic physical or mental health condition. This leniency in screening therefore potentially poses threats to the triads involved in (British) surrogacy arrangements—the commissioning couple, the baby and the surrogate mother herself. The child, if he or she knows who the third parties were, could sue either party for damages if she/he is born with the disease or condition, upon reaching the age of 18. It is therefore in everyone's interests that surrogacy arrangements are not entered into lightly, that all parties are aware of the risks they are taking and ensure that these risks are minimised.

As a result of the perception of risk (as detailed, e.g. in Chap. 10), it was believed that regulation was needed. There were also serious concerns about commercialisation possibilities leading to the commodification of children. It is partly for these reasons that the British Minister for Health at the time commissioned a report in the 1980s and again in the 1990s (Warnock 1984; Brazier et al. 1997) into surrogacy practices. Meanwhile, concerns over the lack of psychosocial screening, particularly for surrogate mothers relinquishing a baby when they may believe they cannot go through with it, were reported (van den Akker 1998). Most British clinics involved in IVF surrogacy called for the need for a regulatory body overseeing a good outcome for all involved in this practice. The agencies too urged for regulation, and since they had the essential experience to demonstrate the particular areas within surrogacy that should be the focus for regulation, it is surprising that it took a long time for input from them to be sought and accepted.

Organisations Involved with Surrogacy Arrangements

In 1998, only eight clinics in the United Kingdom had a licence to carry out treatments involving surrogate arrangements and of those only six had completed two or more surrogate cases in the previous year (van den Akker 1998, 1999). It is estimated that somewhere in the region of 15 clinics now hold an HFEA licence for surrogate treatments. Compared to the approximately 100 clinics licensed to provide assisted conception services such as IVF, DI and ICSI (intracytoplasmic sperm injection), the number of clinics involved in surrogacy is peripheral. Reasons for this are because surrogate arrangements, though increasing, are not common and since such arrangements are costly to the commissioning couples, areas of least affluence are less likely to apply for a licence to carry out IVF or DI surrogacy. Surrogacy is also more time consuming than traditional IVF treatment. Recent research also suggests there is also a lack of evidence for the benefits of some of the many and often costly treatments involved in bringing about assisted conceptions which provide an additional burden at a practice level (Heneghan et al. 2016) particularly to commissioning parents who are already spending huge amounts of money on the arrangement.

Research designed to find out what the clinics and agencies believed to be their responsibilities within surrogate arrangements was non-existent 20 years ago, the picture was nothing like that presented in the United States. In the United Kingdom, surrogate agencies were and still are primarily focused on negotiating contact between prospective surrogate and commissioning couples (van den Akker 1998, 1999). The prospective surrogate and the prospective commissioning mother are typically asked some questions following which they register with the agency. Registration involves a small membership fee, which is paid for as long as the search for a match takes, and sometimes for years after. Agencies use that money to support their cause, put people in contact with one another, provide information and support services and to produce a regular newsletter. The agency tries to match surrogate and commissioning couples by providing a short profile of one to the other. Each can then state if they wish to make contact with their chosen counterpart and this can go on until both parties feel happy with their 'chosen' couple or surrogate.

No psychological testing, social work investigations or medical examinations are needed, and the matching is carried out entirely between the parties involved. Even the agencies, beyond making some recommendations, have little say in the matter. The result is that some individuals may change their mind about the suitability of the chosen couple or surrogate following a subsequent meeting, or in worst-case scenarios this can happen after the mutual agreement has been reached. The agencies provide a sample contract, which each side is asked to read and sign either in the presence of another person of their choosing or a legal representative. In practice, this is unenforceable in law, but in theory it is intended to make both parties feel they have signed something they hope each party will adhere to. Thus, faith is the factor the entire agreement hangs on. A clinic's involvement in surrogacy is different. Clinics explicitly state the prospective commissioning couple is responsible for finding their own surrogate. In the United Kingdom, clinics are not involved in matching people, but only deal with them once they have found their surrogate and try to assist with the surrogate's DI or IVF conception. More detailed information is available from COTS, Surrogacy UK (1993) and, more recently, from Brilliant Beginnings (www.cots.uk.org; SurrogacyUK.org; Brilliant.beginnings.co.uk).

In 1988 COTS was successfully launched as a non-commercial agency. It had few members, but several committed committee members. A decade later it held over 700 members and over 350 surrogate children were born through the involvement of COTS. These numbers are now substantially higher (Crawshaw et al. 2012). COTS' objective is to provide information and support to surrogates and prospective parents about the surrogacy process. COTS runs a 'Triangle' support group which includes a list of counsellors and mediators. All prospective parents and surrogates are required to see a counsellor/mediator, prior to going on the 'active' list. This is to ensure both parties are well informed of the processes involved, and the procedures to be followed to avoid problems later on in the surrogate arrangement. Once a member is on the 'active' list, they receive monthly phone calls from a chat-liner. Chat-liners are all women with first-hand experience as a surrogate or as a commissioning mother, and can therefore provide meaningful advice to the individuals on their lists. They also provide moral support, and in general surrogates and commissioning couples are grateful for this input.

The actual procedure for finding and meeting a couple for the surrogate follows a few strict steps. A surrogate is given some information about a number of couples seeking a surrogate. These details include only basic sociodemographic information, such as the couple's first names, the county of residence, age, profession, weight, height, colouring and details of why they are unable to bear a child. Once the surrogate has decided on a possible couple, she is sent further details about them, excluding their name, address and telephone number. The surrogate must then decide whether she is still interested in that particular couple, and if she thinks they sound like people she would want to help have a baby. If she does, COTS sends the commissioning couple the 'surrogate's details' which are very much the same as those they provided about themselves, again excluding identifiable information about the surrogate. Once they have both seen each other's details, and both parties are still keen to meet,

COTS will release their names and addresses to each other. It is then entirely left to the surrogate and couples themselves to talk on the phone and organise meetings.

COTS (1988) also provides practical advice on a number of issues pertinent to the surrogacy process. This includes medical issues such as the processes involved in gestational (IVF) surrogacy, prenatal testing (e.g. cystic fibrosis) and preconceptual testing (e.g. tests for HIV, sperm count, etc.). Other issues relate to the legal aspects of surrogacy, for example how to apply for a PO (see also Chap. 10) and names of solicitors who have experience of dealing with surrogate cases. Issues of expenses are also outlined; for example, COTS gives guidance on the amount of expenses a surrogate needs and the amount an IVF surrogacy treatment may cost. Last, but not least, they tackle important issues to do with the surrogate child. One factor here concerns the telling of the child how it was conceived, and the telling of a surrogate's own children what their mother is doing with the new baby and why. Useful information about dealing with the media will also be supplied by COTS.

The Surrogate Parenting Centre (SPC) was set up in 1993 by Claire Austin. Her organisation applies similar principles to those providing the foundations for COTS. Their subsidiary called HOPE, like COTS' Triangle, provides the initial contact between surrogate and prospective commissioning couples. In the United Kingdom, the newest agency is Brilliant Beginnings (2013), set up by legal experts Natalie Gamble Associates, and this agency too caters for the needs for intended commissioning and surrogate parents, and they include an emphasis on contracts. In the United States where commercial surrogacy is legal, selection procedures of surrogates are much more stringent than in the United Kingdom with agencies acting as advocates and go-betweens (Ragone 1994).

The two British studies addressing issues of selection, assessment, responsibilities and functions of the organisations involved in surrogacy also addressed the question of how clinics manage to run surrogate programmes in the absence of legal advertising opportunities (van den Akker 1998, 1999). The studies reported that clinics rely on the agencies to help with matching their infertile couples with the agencies surrogates. The additional clinical involvement in gestational surrogacy via assisted conception leads to a number of benefits for intended couples and the surrogates. Gestational arrangements receive more care from the medical and health care teams and surrogate mothers are provided with a certain amount of protection via the HFEA (1990) licence (see Blyth 1994, 1995). Despite this, no psychosocial protection or care is offered, the risks to both parties are substantial and the financial burden carried by the commissioning parents is great. In terms of drawbacks of clinical involvement, this may well outweigh the benefits over genetic surrogacy, although in the latter the commissioning mother will not have a genetic link with the child and the surrogate relinquishes a child with a genetic link which have their own complications (van den Akker 2002, 2007). American commissioning parents are screened for 'parental fitness' and thorough psychological screening by the agencies and medical screening of surrogates by the clinics is carried out (Ragone 1994). In the United Kingdom, no uniformity of screening is evident. More importantly, although the successes seem to outweigh the losses, with a few notable exceptions, the longer-term social and emotional impact of surrogate mother arrangements of the children when they have grown up, the surrogate mothers and commissioning parents is rarely considered. This reflects the poor practice ignoring the long-term mental health of those involved in gamete donation (see e.g. Crawshaw et al. 2016; Scheib et al. 2017; Blyth et al. 2017). Specifically, issues of attachment and bonding in the triads involved are not yet adequately addressed (see Chap. 6).

Extreme Surrogacy

Advertising for services and goods connotes business ware for sale and purchasing, which is neither desirable in human commodification terms, nor legal in, for example, the United Kingdom. Some practices could result in criminal offences, thereby tainting an area of reproduction already marred by much controversy. In Thailand, a 'surrogacy ring' was uncovered where criminals trafficked a number of Vietnamese pregnant and not yet pregnant women for the sole purpose of serving as surrogates (ABC/News 2011). A search in Google of 'embryos for sale' will result in pages of listings, many of these from the United States. One example of such

Table 3.1 Showing an example of a surrogacy 'package' deal

Included	Excluded
The cost of a first-time US-based egg donor	Travel and accommodation for the donor
The cost of the donor agency and donor screening	Travel and accommodation for the parents
Surrogate recruitment and screening	Premium egg donors[a]
IVF medications, embryo transfer	Sperm analysis
Legal needs of the surrogate and donor	Housing for the surrogate for the first seven months of pregnancy
Protocol monitoring	Triple marker and amniocentesis
Sperm freezing	Passport and DNA testing
Two pregnancy tests	Twins ($6000 extra)
Financial compensation to the surrogate	Caesarean section ($750 extra)
Allowance for surrogate mother's food and clothes	If you do not need an egg donor the price drops to $33,900
Third-trimester travel and accommodation for the surrogate	If you want to ship your own embryos the price is only $32,200 plus shipping
Normal delivery	
Gender selection	
Emergency medical for the surrogate	
$39,995 all inclusive[b]	

[a]Defined as speciality donors and donors who have donated before
[b]All-inclusive indicates the costs of a normal surrogacy procedure. In case of specific requests or medical conditions, additional charges will apply

a website shown in Table 3.1 offers 'American egg donors and Mexican Surrogates'. It offers 'package' deals which include/exclude the following:

Advertising a successful business enterprise means operating with an awareness of market demands, and surrogacy companies are just as cunning as non-surrogacy industries, offering additional benefits, guarantees and consumer choices. For example, wealthy Chinese couples are noted in the news (see e.g. Symons 2013a) as:

Using US surrogates, to obtain automatic residency in the United States American companies, like the one offering specific deals above, will offer people what they want, and Chinese want what money can buy. The benefits of US surrogates for the Chinese for example include, avoidance of the one-child policy (now revised), decreased risks of developmental complications, you can 'help' design your own baby, and most importantly, the child can get US citizenship. Under the 14th amendment, any person born on US soil is eligible for citizenship, and an American child will help the parents get a green card. Symons (2013a) reports "If you add in plane tickets and other expenses, for only $300,000 you get two children and the entire family can immigrate to the US."

Since some countries discriminate against single (including the United Kingdom) or gay men to use surrogacy, they too tend to approach cross-border surrogate agencies, which will help them to build their families:

Agencies specifically targeting these commissioning parents;
.... pay between "$125,000-150,000 for a singleton and between $150,000-175,000 for twins". (Growing Generations, an LA based surrogacy agency, cited in Cook 2013a)

Just as there is still overt discrimination of same-sex and single parenting at legislative levels in many countries, there remains a substantial amount of discrimination of disabled parents. Symons (2013b) reported on an Israeli woman disabled from muscular dystrophy. She commissioned a genetic surrogacy arrangement using an Israeli man's sperm. Social services intervened because she was not the legal (non-genetic) parent of the baby, but the woman's lawyer said this decision to put the baby up for adoption was due to the 'perceived inability' of the woman to care for the baby.

Across the world, in countries where surrogacy in some form is legal, advertising for the supply of surrogates is also often possible. However, when surrogacy is operated like a business the quality needs to be assured and delivery upon payment, at least in business, is mandatory. This can pose numerous issues which are considered ethically, morally, socially and psychologically unacceptable, and can have significant consequences. For example, consider a surrogate advertising her services as quoted in BioEdge;

I want to become a surrogate mother. I am 23 with a 2.5 y.o. child.... I have no bad habits. I had no abortions or miscarriages. I am Kazakh, 160cm tall, weight 60kg. Blood type AB+. Physically and psychologically healthy. I have no previous experience of being a surrogate mother. I am ready to move to Almaty or Astana. Legal contract compulsory. Remuneration $20,000 plus 50 thousand tenge ($333) per month during pregnancy. I will consider all offers. (Cook 2013b)

This surrogate, operating outside of a company framework, may not be in a position to defend her rights should problems arise or demands be made which she believes may be contrary to the contract or her personal beliefs of right and wrong. The surrogacy industry is also problematic because the surrogates are not machines. They may not adhere to the contract, renege on relinquishing the baby, ask for more money than was agreed or disagree with the proposed medical procedures during pregnancy, as Symons' (2013c) spine-chilling report shows:

A Connecticut couple offered their American surrogate mother US $10,000 to abort their baby because it had major defects. When the surrogate refused, the couple's lawyer sent a letter demanding that the baby be aborted. 'you are obligated to terminate this pregnancy immediately .... you have squandered precious time.'

Another British surrogate ended up handing one of her 13 surrogate babies from nine pregnancies—the latter triplets—to the commissioning couple when in fact it was her and her partner's baby (Cook 2012). She said:

Paul felt terrible he'd got me pregnant. We discussed whether we should take the baby back. But I couldn't do that to another woman – it would have killed her. If I thought that she did not want him I would have him but I knew 100 per cent that she did.

Illegal sex selective abortions are also common in international surrogacy, including the American and Indian surrogacy industries. Sex selection in India was banned in the 1990s, yet a discrepancy between male and female infant births was still recorded in hospitals in 2013 (Cook 2013c). The practice appears to be driven by commissioning parents themselves seeking surrogacy in India, and half of those questioned in a Centre for Social Research report reported agreeing to sex selection in cases where foetal reduction was necessary. Finally, because no cross-border checks of parental suitability exist, there is nothing preventing criminals commissioning babies. Paedophiles, for example, have been known to commission surrogate babies (see again a report by Cook 2013d).

Counselling

In 1988, the British Infertility Counselling Association (BICA) was set up in the United Kingdom to focus on counselling in infertility and related issues including practical, social and psychological issues and ethical issues and concerns. As noted previously, offering counselling services to intended and surrogate parents became mandatory in the United Kingdom following the 1990 HFEA Act demonstrating a clear need and recognition of the importance attributed to counselling issues in infertility treatment. There is, however, no statutory requirement for its uptake. Research and practice have shown that not all patients will take up counselling when it is offered (Shaw et al. 1988; van den Akker 1999; Payne and van den Akker 2016). In 2008, the HFEA's Code of Practice-licensed centre counsellors were expected to be accredited, a process facilitated through BICA (1988). Infertility counselling offers individuals an opportunity to confidentially reflect upon a number of concerns, including: on private and sometimes painful thoughts and feelings surrounding their infertility; their hopes and fears about the different types of treatment options; and on the likely consequences of their decisions. This is called implications counselling (Jennings 1995). Counselling also provides informational and understanding support, hence the term 'information counselling'. BICA has developed numerous publications specific to different aspects of infertility and treatments, including surrogacy, via its Guidelines for Good Practice in Infertility Counselling. It is currently the only British professional association for fertility counselling and counsellors for all aspects of the infertility journey, its treatment and post successful or unsuccessful outcomes. BICA accredited counsellors are available nationwide via their website.

The amount and type of counselling that may be beneficial to the commissioning individual or couple and the surrogate mother will depend upon the person and the type of surrogacy, as well as the previous reproductive history of both parties (van den Akker 2002). There is little systematic assessment of the impact of counselling upon those involved in surrogacy using it, but it has been described as limited prior to 1988 (Edelmann and Connolly 1986). Satisfaction with counselling services in infertility treatment generally has been described as beneficial, but where patients failed to get pregnant, satisfaction decreased (Donegan 1994). In another study (Smith et al. 2000), it was shown that patient information and understanding is not always adequately processed, which can lead to unrealistic expectations of the success rates. Counselling which incorporates a realistic and reinforced informational component as well as an emotional care model is therefore recommended to improve outcomes. Helping people form not only realistic expectations but also appropriate emotional adjustments throughout the treatment or surrogate motherhood process is then likely to be more effective in the short term. In the long term, coping with a new family brought about using third-party-assisted conception may require some cognitive restructuring (van den Akker 2001a), as will relinquishing a baby (particularly with a genetic link) (van den Akker 2003, 2005a, b). Not succeeding in building a family through surrogacy also has a long-term impact which is often unrecognised and unsupported (van den Akker 2001a, b). Counselling therefore also has a role to play in the immediate and long-term effects of successful and unsuccessful third-party-assisted conception.

Research and reports on infertility and related gynaecologic or obstetric conditions also demonstrate that the (infertility) treatment trajectory, including assisted conceptions in a surrogate motherhood arrangement, is too focused on the technological and medical aspects and too little on the emotional rollercoaster the patients go through (Schmidt 1998; Wingfield et al. 1997; Payne and van den Akker 2016). In that sense, the health care system fails to meet the psychological needs of infertile people and surrogate mothers. Informational, psychological and psychosexual counselling could and should be infused in equal measure to ensure those coming out of assisted conception clinics at the other end, either with a baby in their arms or without, are well equipped to deal with the new challenges they will face as surrogate parents, surrogate mothers or failed surrogate or commissioning couples. Adequate provision with follow-up and an assessment of the efficacy of counselling is necessary to ensure this becomes accessible, tailored and effective for all those involved with assisted conception treatment in surrogate motherhood arrangements, in preparation of their life long mental health following the choices they have made. For those who successfully build a family through surrogate motherhood, additional specific family mediation and support services may be necessary over the lifespan, focusing upon individual development, attachment relationships, communication and disclosure within families, and the long-term impact of the infertility, of the treatment and of the resultant family (Pettle 2003).

Summary

Chapters 1, 2 and this chapter in Part I have shown that the concept of surrogacy is not consistently clearly defined or used; instead, the terminology can be misleading and ambiguous (Smolin 2016). This harsh reality represents an area of developing research, theory and practice which is riddled with inconsistent, uncoordinated and poorly choreographed issues at the individual, social, national, legal and global levels. This chapter has demonstrated that what is necessary is equity in access to help in family building in social contexts which are ready to accept non-traditional families. Future humane multidisciplinary treatment of the parties involved in and resulting from these arrangements is critical, with professionals taking full responsibility for the process and outcomes. Counselling will need to be used increasingly to ensure the parties involved fare well before, during and after these difficult decisions about surrogacy are made, regardless of the short-term outcomes.

References

ABC/News. (2011, February 25). Women freed from 'inhuman' baby ring. http://www.abc.net.au/news/2011-02-25/women-freed-from-inhuman-baby-ring/1956588. Accessed 27 May 2016.

Blyth, E. (1994). "I wanted to be interesting. I wanted to be able to say 'I've done something with my life'": Interviews with surrogate mothers in Britain. *Journal of Reproductive and Infant Psychology, 12*, 189–198.

Blyth, E. (1995). "Not a primrose path": Commissioning parents' experiences of surrogacy arrangements in Britain. *Journal of Reproductive and Infant Psychology, 13*, 185–196.

Blyth, E., Crawshaw, M., Frith, L., & van den Akker, O. (2017). The modern family and the future? Gamete donors' motivations for, expectations and experiences of, registration with UK donor link. *Human Fertility*, 1–11. doi: 10.1080/14647273.2017.1292005.

Brazier, M., Campbell, A., & Golombok, S. (1997). *Surrogacy review for health ministers of current arrangements for payments and regulation*. Report of the review team. Cm 4068. London: Department of Health.

Brilliant Beginnings. (2013). www.vrilliantbeginnings.co.uk. Accessed 25 May 2016.

British Infertility Counselling Association. (1988). http://bica.net/about. Accessed 29 Apr 2016.

Cook, M. (2012, November 11). Surrogate mother of 13 calls it a day. *BioEdge*. http://www.bioedge.org/bioethics/surrogate_mother_of_13_calls_it_a_day/10313

Cook, M. (2013a, October 22). Single dad and surrogate mother: a growing combination in US. *BioEdge*. http://www.bioedge.org/bioethics/single_dad_and_surrogate_mother_a_growing_combination_in_us/10662

Cook, M. (2013b, April 13). Are US media finally discovering the exploitation in India surrogacy? *BioEdge*. http://www.bioedge.org/bioethics/are_us_media_finally_discovering_the_exploitation_in_india_surrogacy/10464

Cook, M. (2013c, August 17). Illegal sex-selective abortions widespread in Indian surrogacy industry. *BioEdge*. http://www.bioedge.org/bioethics/illegal_sex_selective_abortions_widespread_in_indian_surrogacy_industry/10645

Cook, M. (2013d, June 15). Convicted paedophile gets child from surrogate mother in India. *BioEdge*. http://www.bioedge.org/bioethics/convicted_paedophile_gets_child_from_surrogate_mother_in_india/10560

COTS. (1988). Childlessness Overcome Through Surrogacy. www.surrogacy.org.uk. Accessed 25 May 2016.

Crawshaw, M., Blyth, E., & van den Akker, O. (2012). The changing profile of surrogacy in the UK – Implications for policy and practice. *Journal of Social Welfare and Family Law*, 1–11. doi:10.1080/09649069.2012.750478.

Crawshaw, M., Frith, L., van den Akker, O., & Blyth, E. (2016). Voluntary DNA-based information exchange and contact services following donor conception: An analysis of service users' needs. *New Genetics and Society, 35*(4), 372–392.

Donegan, C. (1994). An assessment of the counselling needs of GIFT recipients: A pilot study. *Journal of Reproductive and Infant Psychology, 12*, 127–130.

Edelmann, R. J., & Connolly, K. J. (1986). Psychological aspects of infertility. *British Journal of Medical Psychology, 59*, 209–219.

Heneghan, C., Spencer, E., Bobrovitz, N., et al. (2016). Lack of evidence for interventions offered in UK fertility centres. *British Medical Journal, 355*. doi:10.1136/bmj.i6295.

Human Fertilisation and Embryology Act. (1990). Her Majesty's Stationary Office. London.

Jennings, S. E. (1995). *Infertility counselling*. Oxford: Blackwell Science.

Payne, N., & van den Akker, O. (2016). Infertility network UK survey on the impact of fertility problems. Final report, May.

Pettle, S. (2003). Psychological therapy and counselling with individuals and families after donor conception. Chapter 8. In D. Singer & M. Hunter (Eds.), *Assisted human reproduction: Psychological and ethical dilemmas* (pp. 155–181). London/Philadelphia: Whurr.

Ragone, H. (1994). *Surrogate motherhood: Conception in the heart*. Boulder/Oxford: Westview Press.

Scheib, J., Ruby, A., & Benward, J. (2017). Who requests their sperm donor identity? The first ten years of information releases to adults with open identity donors. *Fertility and Sterility, 107*(2), 483–493.

Schmidt, L. (1998). Infertile couples' assessment of infertility treatment. *Acta Obstetricia et Gynecologica Scandinavica, 6*, 649–653.

Shaw, P., Johnston, M., & Shaw, R. (1988). Counselling needs. Emotional and relationship problems in couples awaiting IVF. *Journal of Psychosomatic Obstetrics and Gynaecology, 9*, 171–180.

Smith, C., Bayley, L., Adhege, J., & van den Akker, O. B. A. (2000). Patient satisfaction with assisted conception services: The role of expectations. *Journal of Reproductive and Infant Psychology, 18*(3), 265.

Smolin, D. (2016). Surrogacy as the sale of children. *Pepperdine Law Review, 43*, 265–311.

Surrogacy UK. (1993). https://www.surrogacyuk.org/. Accessed 25 May 2016.

Symons, X. (2013a, October 5). Chinese couples using American surrogates to get residency. *BioEdge*. http://www.bioedge.org/bioethics/chinese_couples_using_american_surrogates_to_get_residency/10701

Symons, X. (2013b, May 18). Thorny custody case could set precedent for disabled parents. *BioEdge*. http://www.bioedge.org/bioethics/thorny_custody_case_could_set_precedent_for_disabled_parents/10516

Symons, X. (2013c, March 7). Surrogate refuses $10,000 to abort child. *BioEdge*. http://www.bioedge.org/bioethics/surrogate_refuses_10000_to_abort_child/10422

van den Akker, O. B. A. (1998). Functions and responsibilities of organizations dealing with surrogate motherhood in the UK. *Human Fertility, 1*, 10–13.

van den Akker, O. B. A. (1999). Organizational selection and assessment of women entering a surrogacy agreement in the UK. *Human Reproduction, 14*(1), 101–105.

van den Akker, O. B. A. (2001a). Adoption in the age of reproductive technology. *Journal of Reproductive and Infant Psychology, 19*(2), 147–159.

van den Akker, O. (2001b). The acceptable face of parenthood: The relative status of biological and cultural interpretations of offspring in infertility treatment. *Psychology, Evolution and Gender, 3*, 137–153.

van den Akker, O. (2002). *The complete guide to infertility. Diagnosis, treatment, options*. London: Free Association Press.

van den Akker, O. B. A. (2003). Genetic and gestational surrogate mothers' experience of surrogacy. *Journal of Reproductive and Infant Psychology, 21*(2), 145–161.

van den Akker, O. B. A. (2005a). A longitudinal pre pregnancy to post delivery comparison of genetic and gestational surrogate and intended mothers: Confidence and genealogy. *Journal of Psychosomatic Obstetrics and Gynecology, 26*(4), 277–284.

van den Akker, O. B. A. (2005b). Coping, quality of life and psychiatric morbidity in 3 groups of sub-fertile women: Does process or outcome affect psychological functioning? *Patient Education and Counselling, 57*(2), 183–189.

van den Akker, O. (2007). Psychosocial aspects of surrogate motherhood. *Human Reproduction Update, 13*(1), 53–62.

van den Akker, O. (2012). Chapter 10: Overcoming involuntary childlessness and assisted conception. In *Reproductive health psychology* (pp. 162–165). Chichester: Wiley.

Warnock, M. (1984). *A question of life: The Warnock report on human fertility and embryology*. Oxford: Blackwell.

Wingfield, M., Wood, C., Henderson, L., & Wood, R. (1997). Treatment of endometriosis involving a self-help group positively affects patients' perception of care. *Journal of Psychosomatic Obstetrics and Gynaecology, 18*(4), 255–258.

Part 2

Research on Surrogate Mothers, Commissioning Parents, Parenting and the Surrogate Offspring

Explaining behaviours associated with surrogate mothers involves explaining factors associated with commissioning parents and surrogate babies and vice versa. This part therefore provides a detailed account of research specifically on surrogate mothers in Chap. 4 and on commissioning parents in Chap. 5, separation and parenting and surrogate offspring in Chaps. 6 and 7 respectively. Surrogate motherhood occurs within triads and therefore discussing one without touching upon the other is neither sensible nor always possible or desirable, hence their placement within this second part of this monograph. Research on surrogate mothers is not as prolific as might be expected. Surrogate mothers carry a huge responsibility to themselves, the unborn baby and the commissioning parent(s). They also put themselves at a substantial physical, social and psychological risk and may be exploited and stigmatised (van den Akker 2015; and see Chap. 4). The journey through a surrogate motherhood arrangement for commissioning parents too is riddled with uncertainty, risks and potential stigma due to their infertility and their non-traditional route to parenthood. These issues are discussed in detail in Chap. 5. Chapter 6 covers research and theory on separation and parenting, because surrogate mothers need to adjust to the loss of the baby they conceived, carried and nurtured successfully to term and commissioning parents need to adjust and adapt to parenting a non-gestational and possibly non-genetic baby. The research shows that parenthood and particularly motherhood are concepts that were relatively certain over the centuries. Increased use of third-party-assisted conception and alternative types of parenthood has led to uncertainty in genetic and gestational parenthood. This uncertainty has resulted in a very slow shift in traditional attitudes and has led to questioning the morality of surrogate mothers and the parenting motivations and abilities of commissioning parents. Importantly the uncertainty in parenthood has focused upon the welfare of the resultant child, and these issues relevant to offspring brought about via surrogate motherhood arrangements are covered in Chap. 7.

Surrogate motherhood or surrogate 'kinships' represent a far cry from traditional kinship ideology as represented in much of the Western world. Assisted conception using third-party gametes or embryos in IVF treatment or in surrogates is increasingly common and fulfils a population need of massive proportions, way beyond the needs of the medically infertile. This is because many more diverse families now also seek assisted conception services for social infertility, including single or older women and men and same-sex couples, contributing to the move away from what was known as traditional family structures. These social or lifestyle uses of assisted reproduction for involuntary childlessness represent an increasing proportion of treatments, shifting what was once a heterosexual two-parent family norm. Despite this shift in non-normative family building, national statistics which help to predict future expenditure do not always accurately reflect the consequences of behavioural and lifestyle changes of individuals within populations including accurate birth registrations. Furthermore, the decreasing fertility prospects as a result of lifestyle choices such as delayed childbearing, solo or same-sex parenting mean that many of those not in a medical or social position to conceive without a third party will be requiring assistance to build their families. In the United Kingdom, health care resources may fund these needs for some who request it, but not as frequently for same-sex or single individuals (HFEA 2014). Records show the NHS funds around 40% of IVF treatment cycles, with the remaining 60% privately funded (HFEA 2012). Similarly, for treatments using gamete donation the NHS funds even less. Women in same-sex partnerships and without a partner are much less likely to receive NHS funding for their treatment (HFEA 2014, p. 20). In areas where funding is not provided, health inequalities determine who has and who does not have treatment to overcome involuntary childlessness for social and lifestyle or medical reasons. With the increasing national uptake and international commercialisation of gamete, embryo and surrogate services, further amplification of inequalities of family building possibilities develops and surrogate mothers are party to this developing trend.

References

HFEA. (2012). Fertility treatment in 2012: Trends and figures. www.hfea.gov.uk/docs/FertilityTreatment2012TrendsFigures.PDF

HFEA. (2014). http://www.hfea.gov.uk/docs/Egg_and_sperm_donation_in_the_UK_2012-2013.pdf. Accessed 3 May 2016.

van den Akker, O. B. A. (2015). Emotional and psychosocial risk associated with fertility treatment. In R. Mathur (Ed.), *Reducing risk in fertility treatment*. London: Springer. Science + Media.

4

Surrogate Mothers

Surrogate motherhood takes place in a sociocultural context which still measures differences in parenting in relation to traditional standards. This chapter shows the multiple dilemmas affecting surrogate motherhood practices. Screening, for example, is indicated because of the potential harm the surrogate mother's gestation and/or genetic contribution can do to the unborn child, but this may compromise her autonomy; surrogate mothers de-emphasise the importance of genetic and gestational kinship and not to attach to the foetus, but this can have consequences for the baby. Surrogate mothers also experience stigma, psychological morbidity and medical risks, but support and information to protect surrogates from harm and to develop the long-term cognitive adjustments necessary may be non-existent in research, policy and practice.

Surrogate mothers are intimately involved with the commissioning recipients' desire for a baby; more intimately than, for example, gamete and embryo donors, although they also have some commonalities with them. Surrogate mothers are not only involved in donating gametes if genetic surrogacy is used, but they are instrumental in the gestation of the embryonic cells into foetal and baby status. They experience the growth and movement of the foetus and may form attachments to the baby in vivo, which may be stronger or weaker or the same when

© The Author(s) 2017
Olga B.A. van den Akker, *Surrogate Motherhood Families*,
DOI 10.1007/978-3-319-60453-4_4

their own, or when anonymous or known donor gametes or embryos are used. The issues involved in the donation and receipt of gametes or embryos deserve a book in their own right, but it is worth noting here that some embryo donors have to make difficult decisions regarding their embryos which they sometimes see as 'persons' or babies. When 'surplus embryos' (embryos created for a couple's fertility treatment but not used) need to be discarded or donated to research or another couple (Mansour et al. 2014), including potential couples using surrogacy, they are faced with the possibility that the recipients will be successful—even when they themselves may not have been. The decisions and difficulty depend upon a number of factors that change as the couples' own journey through the infertility treatment maze changes (De Lacey 2013). In a recent study (Bruno et al. 2016), the decision to stop cryopreservation of surplus embryos was more frequent if the embryo was represented as a child, whereas patients more frequently donated to research if they saw the embryo as a project. Finally, if they attributed personhood to the embryo they were more likely to donate their embryos to another couple, and donation to another couple was also 10 times more frequent if they had used gamete donation themselves. Only half of the participants in this study found it easy to make a decision (Bruno et al. 2016).

Like gamete and embryo donors, surrogate mothers may attach to the foetus/baby during pregnancy, and genetic surrogates have an additional genetic link to the baby they carry for a recipient couple. However, unlike gamete or embryo donors, they can represent the baby only as a person, not just a collection of cells or a 'project'. A surrogate mother's input therefore is risky as she is responsible for the lifeline of the conceptus which depends upon her. It may also be genetically related to her, which puts her at a potential additional emotional risk. Surrogates who have formed strong attachments to the baby have experienced grief reactions post-delivery after relinquishment of the baby (van den Akker 2003, 2005; Reame 1989), although this is not always reported. Research into the recipients of donated gametes or embryos including surrogate babies is increasing and has highlighted its benefits. Nevertheless, a number of concerns for the resultant offspring, including their right to know their genetic donor or carrier, are also reported (van den Akker 2006, 2016). Gamete donors themselves are now registering themselves to be available to donor conceived (DC) offspring in case they are searching for them (Blyth et al. 2017). Research into the long-term effects of gamete and embryo donation and surrogacy is still in its infancy. These—third-party-conceived children's—rights may have implications for surrogate mothers long into the future, as is currently evidenced for those conceived via gamete donation and individuals who made the donations (van den Akker et al. 2015; Crawshaw et al. 2016; Blyth et al. 2017).

Suitability Assessment in Surrogate Motherhood

The early reports of women becoming genetic surrogate mothers in the United States were received with dismay across many states. According to Parker (1984, p. 21), 'it has sent shock waves through various quarters.' Genetic surrogacy is usually carried out using artificial insemination by donor (AID) which has been practised for many years in infertile couples and although not devoid of stigma either (couples were advised to keep this a secret) genetic surrogacy was seen as a step too far because here a couple choose the biological mother of the husband's child. The husband stays with his wife who adopts or applies for legal parental status of the child and the surrogate mother relinquishes her biological child to the couple. Since medical intervention is used to bring these arranged pregnancies about, medical ethical principles are applied just like they would in any other medical intervention. These refer to autonomy, non-maleficence, beneficence and justice (see also Part III). The reason why these important principles need to be debated in surrogate motherhood is because there may be instances where the balance of one or more of these principles may be compromised. They are also given as reasons to screen potential parties involved in surrogate motherhood arrangements (Parker 1984), much like happens in adoption.

However, the controversy did not stop there; commissioned parenthood and relinquishing parenthood suitability were debated. Since the process was relatively new, commissioning and relinquishing parenthood entered uncharted territory. In traditional conceptions no policing of those who cause a pregnancy or those who gestate a pregnancy exists, and children are known to thrive in situations which were at one time regarded as suboptimal (Masten et al. 1990). Nevertheless, threats to children can be found within any family type (heterosexual, single, same sex, step-parents, etc.) traditionally conceived or via assisted conception or adoption. There is, however, one critical difference between traditionally conceived and adoptive parents, and the parents involved in surrogacy—surrogate-born children are commissioned.

From the surrogate mothers' perspective screening is important. A genetic or gestational surrogate mother with a substantial smoking, drug or alcohol problem or a genetic surrogate mother with a history of major psychiatric or other medical illness may not be a suitable surrogate if the evidence for genetic transmission of the condition is likely, as this will shape the constitution of the unborn baby (Egliston et al. 2007). The intrauterine environment may be compromised by numerous other negative lifestyle factors in both types of surrogate mother, and a compromised neonatal environment can have effects well into the surrogate offspring's adulthood (Ombelet et al. 2005). However, the ethical principles of autonomy of the surrogate mother may be compromised by mental and physical health and lifestyle screening. In 1984, Parker (p. 24) argued that psychiatric screening of surrogates (and intended parent(s)) was justified even if it interferes with their autonomy 'to protect incompetent applicants as well an unconceived and unborn child'. These points involve a thin balancing act between autonomy and intervention to protect everyone within the triad. Intuitively it seems wrong to remove autonomy through assessment of suitability to be a surrogate mother. In practice, there may be a stronger case for it.

Surrogates may be limited in their ability to make good decisions based on sound judgement and well-understood information (see section below on informed consent). A surrogate may also be at risk for postnatal depression and not be aware that having another pregnancy could trigger another occurrence of this devastating illness. A surrogate may think about the attraction of the income and not consider the implications of relinquishing a baby which may be genetically related to her and will be gestated by her. She may not be aware that she may bond with the foetus in pregnancy, or not feel comfortable relinquishing the baby to the couple or individual commissioning the pregnancy. If she has a husband, he may have issues with the new and intrusive attentions of the commissioning couple and feel sidelined by the attention upon her. Her own immediate and extended family may resent her for the loss of a potential family member to another family. In the United Kingdom, counselling services address suitability assessment and information provision (see e.g. the British Infertility Counselling Association (BICA, Chapter 3)), via BICA accredited counsellors. However, no one can predict the outcome with or without counselling although one can argue that at least every attempt will have been made if counselling is taken up to discuss the range of emotions and possibilities for a successful outcome.

Effects of Differences in Surrogate Motherhood

Kim Cotton, Britain's first surrogate mother, became a genetic surrogate for an anonymous American couple in 1985, which resulted in a legal battle for the couple to take the baby out of the country. The American brokering company did nothing to ease the transfer of rights of the child and there was a substantial amount of negative press associated with this case. In essence there were two main issues; firstly, the arrangement was anonymous, so the surrogate mother had no idea what these parents to her genetic baby would be like. Secondly, since it was a genetic surrogacy, the surrogate's motivations were questioned—how could a genetic mother relinquish her baby to an unknown couple? The case fuelled negativity in the media about this whole new concept of surrogate motherhood. Soon after this case came to the foreground the first legislation on surrogacy—'the Surrogacy Act'—was passed, which pointedly banned commercial surrogacy from British soil. Cotton found it a deeply emotional experience and vowed never to become a genetic surrogate again. She published her first book about the experience (Cotton 1985).

In addition to writing about her experiences, Cotton started 'Childlessness Overcome Through Surrogacy' (COTS), an organisation designed to bring surrogate and commissioning couples together and offer information and support to both parties. Within this supportive network she became a second time around gestational surrogate mother of twins for a couple she got to know well through COTS. This pregnancy, though difficult because it was a twin pregnancy, was easier because she had an ongoing relationship with the commissioning intended couple and there was no genetic link with her. She published another book to balance her first published experience of genetic surrogacy against this gestational surrogacy (Cotton 1992). In this book, the language used about the pregnancy is more detached than in 1985; for example, she 'felt as if drowning under a ton of weight'; 'the boy was jammed under my rib cage'; 'the girl was deep in the groin'; 'I also had pains in my groin, caused by the sheer weight of the bulge I was carrying.' It reflects her lack of attachment to the twins, helped by the lack of genetic link and the constant reminder and presence of the commissioning intended couple reinforcing who the parents of these babies were.

As with Kim Cotton, there has been a shift in type of surrogacy used in latter decades, in tandem with increasingly sophisticated and successful IVF techniques. Now many more gestational as opposed to the less technologically advanced genetic surrogacies are commissioned worldwide (Henaghan 2013). The Surrogacy Act allowed UK clinics to carry out altruistic non-commercial surrogacy without restrictions provided they followed the recommendations made. This, along with slowly shifting public attitudes and increasingly sophisticated IVF techniques, has led to this change in frequency of gestational surrogacy in the United Kingdom and elsewhere.

Contracting a Pregnancy with a Surrogate Mother

Early reports from America of genetic/traditional surrogacy were highlighted by the Baby M case. The surrogate mother refused to give up the baby to the commissioning parents and applied for custody (Corea 1985). The surrogate mother in this case used her own genetic material, and changed her mind after the arrangement. It highlights the potential difficulties some surrogate mothers have in giving up a child that is much like their own child would be. It also highlights the potential harm to the surrogate mother if pre-birth orders are granted in favour of the commissioning parents at a time when the surrogate mother has no idea yet on how she will feel about the baby and how emotionally bonded and attached she may be during and following the pregnancy. Pre-birth orders are heralded by some (e.g. Hinson and McBrien 2011; Horsey 2015) as 'straightforward' solutions to potential surrogacy arrangements where the surrogate reneges on the arrangement. However, these are far from straightforward solutions.

Equally concerning for the surrogate mother and baby is the reverse scenario where the intended commissioning parent(s) may be bound by a pre-birth order but they change their mind, as was the case, for example, with baby Gammy. The baby Gammy case concerns a surrogate twin birth where one of the twins was healthy and the other was a twin with Down's syndrome. The commissioning couple did not want the Down's twin. They separated the twins and took the healthy twin and left baby Gammy with the surrogate mother. It would not be in the best interests of that child to have been forced upon its intended parents although in this case the entire situation was appalling for all parties concerned—the twins, the surrogate mother who ended up raising the affected twin and the commissioning parents who were not fit to become parents. Becoming a surrogate mother, therefore, may not always be as easy as it seems, and pre-birth orders are clearly not a viable solution if the best interests of the child are paramount as was befittingly recognised in adoption half a century ago.

Informed Consent

Research has questioned the ethics of informed consent in surrogate motherhood arrangements. Can surrogate mothers truly consent to the surrogate pregnancy with full information of the process and the implications at their fingertips? The potential barriers to fully informed consenting include the following:

1. Do surrogate mothers get their information from the media or agents?
2. Is the information unbiased?
3. Are surrogate mothers tempted by the perceived benefits as opposed to the costs of becoming a surrogate mother?

4. When recruited by friends or family, or once the couples get to know each other and a desire to help or the feeling of pressure to help emerges, is unbiased informed consent possible when an emotional element clouds this process?

5. In genetic surrogacy it is unlikely that the surrogate mother is aware of the consequences of relinquishing a genetically related baby.

6. In gestational surrogacy she is unlikely to be fully appraised of the potential short- and long-term risks of the IVF process.

7. Are surrogate mothers prepared to face the long-term consequences of health problems in a surrogate baby which can be traced to the uterine environment to which the surrogate mother exposed the developing foetus?

Unbiased informed consent has not been studied sufficiently in relation to surrogate mothers. The lure of money out of poverty or for something 'extra' may influence the decision, or the need to find a way to redeem themselves in relation to their own reproductive history. Emotions may therefore interfere with—and bias—even informed consent.

Characteristics of a Surrogate Mother

Since surrogate motherhood arrangements involve unorthodox pregnancies, it is important to know what differentiates individuals going into such arrangements or what it takes to be a surrogate, what their motivations are and how they fare through a surrogate arrangement. Many negative portrayals have been released through the media, and in some cases they have been correctly described (van den Akker et al. 2016a, b). However, those reported in the media consist of only a small proportion of the many women who are involved in surrogacy and tend to involve celebrity cases or accounts from individuals for whom the arrangement went horribly wrong. It is not surprising that most people who have something to reveal to the media do so because their motivations may be fuelled by experiences of unreasonable or unlikely gains or losses. There are many more surrogates with relatively uneventful experiences who have not been characterised by the media. It is perhaps these women who should provide the answers to the questions posed earlier, not extreme cases.

Personality Profile

Unlike the relatively frequent media reporting on surrogate motherhood, research into surrogate mothers' characteristics including aspects of their psychological profiles is relatively sparse. An understanding of a surrogate mother's personal characteristics is important to help in the determination of who may be suitable. Understanding something about a genetic surrogate mother's (or gamete/embryo donor) psychological health is also important in screening of mental health conditions with genetic contributions to the child, a process routinely carried out in the United States, and not at all in countries like the United Kingdom.

However, the studies which have been carried out have provided some of the evidence needed to determine what characterises a surrogate mother. Franks (1981) published the first systematic report on a sample of 10 surrogate mothers' psychiatric characteristics, which was followed by a report by Hanifin (1987) and van den Akker (2003) and more recently by Pizitz et al. (2013). All American research used the Minnesota Multiphasic Personality Inventory (MMPI) and MMPI-2, but samples were small for the huge numbers of items on these scales (10–43). van den Akker (2003) in the United Kingdom used the Eysenck Personality Questionnaire. All studies reported a lack of psychopathology in the surrogate mothers studied. van den Akker (2003) assessed the personality profiles of 24 surrogate mothers: 11 were gestational surrogates, 13 genetic surrogate mothers; otherwise, there were no sociodemographic differences between them. Results showed that genetic surrogate mothers were no different from their gestational counterparts on any of the personality and general health scales or on quality of life domains, and did not differ from other normative population profiles.

Motivations to Become a Surrogate Mother

In 1994, Snowdon in a paper entitled 'What makes a mother' reported on interviews with women donating eggs and women who were surrogates. Two surrogate mothers were interviewed; one became a surrogate mother because of the financial benefits associated with it; the second surrogate mother was motivated to get involved as a result of watching a programme disputing custody between a surrogate mother and a commissioning couple. Her motivation was therefore altruistic, whereas the other surrogate mother did it primarily for financial gain. Since these motivations are disparate, characterising a surrogate mother is therefore likely to be difficult. In an attempt to understand British surrogate mothers' characteristics and motivations, van den Akker (2003) reported the majority of non-commercial British surrogate mothers had altruistic reasons for becoming a surrogate although a few did this for financial gain from the expenses they received. This is surprising since British legislation stipulates that only non-commercial surrogate motherhood is acceptable, and indicates that the interpretation of 'reasonable expenses' is relatively fluid.

Surrogacy for financial gain in commercial surrogacy is particularly notable in low-resource countries operating cross-border surrogacy arrangements (see Chap. 8). Pande (2010) describes Indian surrogacy as a new and unusual form of women's labour, where women become surrogates only because they need the money. However, there are reports of other surrogate mothers doing it for purely altruistic reasons, even in countries where surrogate motherhood is commercially driven. In America, commercial surrogate mothers have been described as wanting to give the gift of life (Ragone 1994), although they are paid for their services. Some of Ragone's surrogates also became surrogate mothers because they knew someone who was infertile; because they understood what joys they themselves have experienced holding their own babies following delivery; or they may have wanted to correct some previous 'wrong' like an abortion or a miscarriage. For example, about a quarter of women (26%) considering becoming surrogate mothers interviewed by Ragone (1994) had previously undergone an abortion and another 9% had placed a child up for adoption. These previous losses in women who become surrogates are also described by Parker (1983) and Ciccarelli and Beckman (2005). In van den Akker (2003) 3/11 gestational surrogates (27%) and no genetic surrogate mothers had previously had a social termination of pregnancy; 4/11 (36%) gestational and 4/13 (30%) genetic surrogate mothers had experienced a previous miscarriage; all had children of their own and genetic surrogate mothers had carried significantly more surrogate babies than gestational surrogate mothers. Two actually said they became surrogates to deal with their own previous losses of babies. Another proportion did this for reasons of self-interest, including developing a higher self-esteem and empowerment. Surrogate mothers reported they experienced a sense of importance in relation to the people they were matched up with and their own family and friends. Their life took on a new meaning; they became involved in medical practices they had not been involved with prior to the arrangement; they met people they would otherwise not have met; they were valued and in some cases revered.

Some information is consistently reported; most surrogates believed surrogacy takes a special type of person, and for them, this event was one of the most important in their life (Blyth 1994; van den Akker 2003). They also reported that not everyone is suitable to be a surrogate. Most surrogates seem to know if they can or cannot do genetic (straight) surrogacy, that is, if they can or cannot relinquish a baby which is genetically theirs. However, these are snapshots in time and do not provide the long-term picture necessary to be certain that this is stable over a longer period of time. Some surrogate mothers may be misleading themselves and choose genetic or gestational options on the basis of the medical interventions involved, or anticipated differences in money paid to them. Others may be too young to understand the consequences of their actions or may think they do and then regret their decisions later upon the time of relinquishment or even later in life when it is too late to do anything about it. However, most surrogates report their experience as providing them with an enormous amount of satisfaction at having been instrumental in fulfilling a dream of a most desired baby for their commissioning couple (Blyth 1994; van den Akker 2003). Some were so happy with their role as surrogate mother that they were willing to do it a second or third time for the same or other couples; others were happy that they met commissioning couples and became their friends (van den Akker 2003, 2005).

Parker (1983) reported on 125 women who had applied to a surrogate motherhood programme to become surrogates. Although these women were not all familiar with the experience of a surrogate pregnancy, their motivations were the same as those reported a decade later by Blyth (1994) and two decades later by van den Akker (2003). These robust motivations over time consisted of:

- a desire and need for money
- enjoyment they thought they would obtain from the pregnancy
- desire to be pregnant
- the advantages of relinquishing the baby outweighed those of keeping it
- it would help them solve unresolved previous experiences such as abortion or other foetal loss

Hanifin (1987) studied 89 surrogate mothers and produced similar results:

- enjoyment of a pregnant state
- desire to be pregnant again
- empathy for childless couples
- a desire to do something unusual with their life
- financial gain
- compensation for a previous loss in their own personal reproductive career

The anthropologist Helena Ragone (1994) has written detailed qualitative accounts of her observations of and discussions with surrogate mothers. Ragone insists that one cannot even begin to look at surrogacy without examining the 'role of women's work and the separation of domestic work from the public sphere' (p. 51). She argues that the relevance of the separation of the home and the workplace, since the days of the Industrial Revolution, marked the separation of the domestic and public spheres. Surrogacy, she argues, bridges that gap between the public and domestic arenas. Ragone's (1994, 1996) work reveals that American surrogates see surrogacy as a vocation allowing them to express and fulfil themselves. The terms used by gestational surrogates to describe their views of the surrogate baby following delivery revealed a certain amount of detachment as a 'mother' to this child. Terms used included: I was like a 'hotel'; a 'cow'; 'It wasn't mine'; 'I could not relate to the baby as mine'; 'I am only carrying the baby'; 'I did not want to care for them.'

More recently, a very different kind of surrogate mother has been studied. These surrogate mothers are from countries where surrogacy is operated unapologetically within a multi-million-dollar industry (BBC 4, 2013; see also Chap. 8). Surrogate mothers are recruited, controlled and used to produce babies for considerably wealthier couples abroad. The picture is therefore much more amplified than what was seen in the United Kingdom (economic and educational disparities between the surrogate mothers and intended parents, van den Akker 2003, 2005) and America, where these differences are in turn more pronounced than in the United Kingdom. In developing countries such as India, the socioeconomic difference is huge. Karandikar et al. (2014) studied 15 illiterate, Indian surrogate mothers, living in relative poverty. They had been a surrogate mother for a second or third time when they were interviewed, and all said financial motivation drew them to be a surrogate mother. The individual stories at times indicate desperation and continuing onto a second and even third pregnancy despite not really feeling up to it, just to continue to make the money they needed. Others clearly stated they felt they had no choice. In Karandikar et al.'s (2014) study coercion from the husbands to be a surrogate or oocyte donor was also evident. In 10 of the 15 women studied, the husbands made the decision for the women to become surrogates. This is discussed in depth in Part III.

Empowerment and Knowledge

As mentioned above, surrogates' motivations to become surrogate mothers related to self-interests, including having feelings of empowerment. These feelings are in part due to the surrogacy process itself since surrogate mothers tend to be seen as having the upper hand in the arrangement and also in part due to the new social position they find themselves in. Many surrogates commented on being proud to have new professional friends they would not normally have encountered in social situations. Others felt that the sky was now more limitless, they felt like they could climb a mountain, or gain a university degree, showing how the surrogate arrangement had empowered them in many different ways (van den Akker 2003, 2005; Blyth 1994).

British surrogate mothers do not necessarily have a lot of information about a process that can change their lives as much as it does the commissioning couple. Survey evidence has shown that most of the information they have about surrogacy comes from the surrogate agency, the clinic or their own research (van den Akker 2003). Most of the knowledge they had was medical and practical, but their understanding of legal, social and psychological aspects of surrogacy was minimal. However, this limited information from selected sources did not deter them from engaging fully with the surrogate mother process and most became involved well within a year of contacting the agencies (van den Akker 2003).

The Surrogate Mother/Commissioning Parent Relationship

Both happy and unhappy recollections about surrogate/intended couples' relationships are reported. Ragone's (1994) interviews revealed enduring solidarity between the women, but also negative experiences between surrogate mothers and their commissioning parents. Some responses included 'I felt they had been my friends, but after they got what they wanted, they weren't.' This is not uncommon, and has also been reported in the United Kingdom, where several surrogates had felt let down once the couple had their baby (van den Akker 2003, 2005). In many cases, contact becomes irregular and cooler, or stops altogether. Not all surrogates (and their partners) were prepared for the sudden loss of attention and friendship.

The Psychosocial Context of Surrogate Motherhood

Stigma, separation between the surrogate mother and her own children for the duration of the pregnancy in some clinics and deception are also reported. In some international surrogate-motherhood cases, the ostracisation has led to the surrogate mother and her family uprooting after the completion of the surrogate arrangement and moving to another village (Karandikar et al. 2014). Social stigma resulting from having been a surrogate mother may therefore last a lifetime and affect a surrogate mother's extended family. The loss of the mothers' care and proximity to her own growing family has not yet been studied sufficiently. In a British study, stigma and time away from a surrogate's own children, as well as deception by surrogates about their pregnancy, have been reported (van den Akker 2003). Surrogates felt bad that they had to put their own children's needs to the side because they felt unwell during the pregnancy. Others were concerned about society's ignorance about surrogacy and 'having to' pretend their baby had suffered sudden infant death syndrome (SIDS) showing extreme forms of deception. Feelings of guilt at having accepted the expenses and psychological distress including post-natal depression (PND), although infrequent, are also reported in surrogates (van den Akker 2003, 2005; Söderström-Anttila et al. 2002), resulting in potential loss of family functioning. Loss of earnings following time off for post-natal depression or other ill health during or following the pregnancy also has potential economic implications for the surrogate mothers' own future.

Concerns About Being a Surrogate Mother

Not much research has asked surrogate mothers if they have any concerns or worries about the surrogate arrangement or the baby. In another study (van den Akker 2003) genetic and gestational surrogate mothers were separately asked what their worries and what their partners' (if they had one) worries were. Thirty per cent of gestational surrogate mothers and 60% of their partners, and 15% of genetic surrogate mothers had some fears the intended mother may not take the baby. All genetic surrogates said a genetic link was not important to them, hence their willingness to be a genetic surrogate mother. In contrast, 8/11 gestational surrogates believed it was important, hence their partaking in gestational surrogacy. Genetic surrogate mothers did fear (15%) that they would be unable to relinquish the baby—as did 22% of their partners, whereas only 10% of gestational surrogates and none of their partners were concerned about

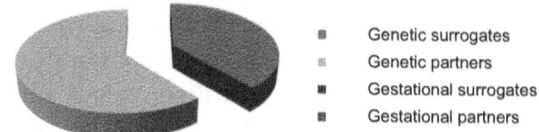

Fig. 4.1 Concerns about not being able to cope with their own emotions

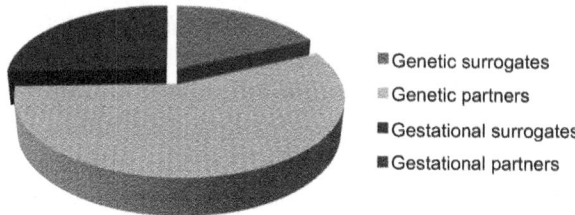

Fig. 4.2 Concerns about the feelings of their own children

this. A further 7% of genetic surrogates—and 11% of their partners—were concerned about not being able to cope with their own emotions whereas none of the gestational surrogates or their partners worried about this (see Fig. 4.1).

In contrast, 10% gestational surrogate mothers—and 40% of their partners—were concerned about their health compared to only 7% and 22% of genetic surrogate mothers and their partners respectively. Finally, as is shown in Fig. 4.2, 10% of gestational and 7% of genetic surrogate mothers—and 22% of genetic surrogates' partners—had some concerns about the feelings of their own existing children in relation to the arrangement.

Maternity Care

The surrogate takes on the risks and burden of pregnancy for another couple or person. Since surrogacy can be carried out without the involvement of medical procedures as in genetic surrogacy using (commissioning father or donor), insemination clinicians and midwifery and health visitor staff may not even be aware the pregnant woman is a surrogate. Once a couple and a surrogate leave the fertility clinic the surrogate mother may be transferred to routine care for the duration of the pregnancy, the delivery and in the post-partum. Accurate records of surrogate births are therefore not possible (Crawshaw et al. 2012). Where it is known that a woman presents with a surrogate pregnancy, the commissioning couple may want to be involved or in charge of the care and decisions to be made during the course of the pregnancy and the delivery. The commissioning couple and medical and health care professionals involved with the care or those who initiated the arrangement must consider the welfare and potential consequences to the surrogate as well as the baby (van den Akker 2015). In some American states that allow for commercial surrogacy, commissioning couples pay all the surrogate mothers' medical, hospital and legal fees involved. Some contracts also include six weeks post-delivery medical expenses and up to six months' cover for treatment of emotional or mental conditions following the date of termination of the pregnancy. Surrogate contracts also typically prohibit the surrogate mother from smoking, drinking alcohol and taking illegal drugs (Charo 1988) and she may, under the conditions of a contract, have to undergo amniocentesis or a Caesarean section (C-section) delivery (Andrews 1982) or any other requests for prenatal care (Charo 1988). The surrogate mother will ideally have undergone some preconception counselling and taken up health behaviours ahead of the surrogate pregnancy. No research exists on the preconception health behaviours of surrogate mothers. No research exists documenting the health behaviours of surrogates during a pregnancy and none exists considering her welfare post-natally, beyond the first six months of pregnancy (van den Akker 2005). These are serious omissions of care and of research foci (van den Akker 2013).

Genetic surrogates are particularly isolated from medical and health care opportunities as little (if DI is used in a clinic) or no intervention (if DI is performed outside of a clinic, e.g. in the surrogate's home) is needed. The gestational surrogate, on the other hand, proportionally receives a considerable amount of attention prior to embryo transfer and implantation. She may continue to receive health care attention during the pregnancy for monitoring purposes and during the delivery. However, post-delivery, she too is removed from all care. It is that time after the pregnancy and delivery when she returns to normal life that the full implications of the arrangement may hit her. Research and practice is beginning to recognise that surrogate mothers may need additional psychological or social support during pregnancy, delivery and the postpartum period, alongside the psychological assessment needs now recognised for all pregnant women (McKauley et al. 2011).

Maternity care provision is different in different countries depending upon many factors, including economic (state funded vs private care), cultural and attitudinal factors. In some countries where cross-border surrogacy was prolific, such as Cambodia, Thailand and India (discussed in more detail in Chap. 8), maternity care is provided to surrogate mothers to the extent that it is somewhat prescriptive. An extreme example is a well-known clinic in Anand, where women were housed in the clinic in a hostel-type dormitory situation for the duration of the pregnancy. The women are fed, medicated and do not have much exercise. They are not meant to have sex with their husbands and spend much of their time in a semi-controlled environment. Researchers of Indian surrogates including Pande (2009), Bromfield and Smith-Rotabi (2014) and Karandikar et al. (2014) have commented on the type of care women receive during the pregnancy. Care is withdrawn soon after the baby is born, relinquished and the women are sent home. The consequence is that no one monitors the post-natal health of surrogate mothers in the immediate and longer term. A similar lack of continuity of care is evident in the West, with surrogates being left to cope with their feelings in relative isolation because showing their emotions may perpetuate a sign of weakness and judgement on their actions.

Obstetric Factors

Apart from the emotional wellbeing of surrogates, the medical care of the physical aspects of the pregnancy are underreported and understudied. Media reports of surrogates dying following a delivery exist (Bionews 2005), but few studies report on the physical welfare of surrogates. Fischer and Gillman (1991) reported an equivalence of somatic symptoms reported during pregnancy in surrogate and non-surrogate pregnant women. They also reported more positive attitudes towards pregnancy in their non-surrogate group. In Karandikar et al.'s (2014) study, all surrogates underwent C-sections, some multiple C-sections for multiple surrogate pregnancies. Pain is controlled via painkillers, but since there are additional risks to the mother and infant associated with C-sections (Clark and Silver 2011; Silver 2012) it is critical that the need for C-sections is re-evaluated. Interestingly IVF or gestational surrogacy using donate oocytes (oocytes from a donor or commissioning mother to the surrogate) puts the surrogate mother at higher risks of placental disorders and gestational hypertension and pre-eclampsia. Gestational surrogates, like oocyte donor treated cycles, are at increased risk of poorer neonatal outcomes because this is increased in oocyte donation pregnancies compared to non-oocyte donation IVF pregnancies. Even poorer outcomes are reported for twin pregnancies (Savasi et al. 2016).

In terms of the surrogate mothers' clinical care, an obstetrician and midwifery team has to consider the best interests of the foetus and baby, the pregnant and post-natal surrogate and the intended parents, particularly in countries where litigation is prolific. Accurate obstetric histories must be taken into account in a surrogates' surrogate pregnancy and valid informed consent of the immediate and long-term medical, social and psychological consequences is critical (Burrell and Edozien 2014). To date, the majority of reports of surrogate pregnancies indicate problem-free pregnancies and deliveries (Corson et al. 1998; Parkinson et al. 1999; Brinsden 2003). Parkinson et al. (1999), for example, reported on a number of medical risks gestational surrogates were affected by. These risks included pregnancy-induced hypertension and bleeding in the final phase of the pregnancy, although this was four to five times lower in gestational surrogates compared to IVF infertile patients regardless of multiple or single pregnancies. This lower rate may be in part because most surrogate mothers have had successful pregnancies before and in part because they tend to be relatively young compared to the rest of the childbearing population. However, the incidence of C-sections is high in gestational surrogates (21.3% for singleton; 56.3% for multiple gestations) and obstetric complications are known (Duffy et al. 2005). In Duffy et al.'s (2005) report two out of nine surrogates had a post-partum hysterectomy following a placenta accrete and a uterine rupture. These incidents need to be monitored, particularly when multiple embryos are transferred, as they should form part of the 'success' or 'failure' rates recorded in IVF clinics. The author is not aware of clinics routinely monitoring the outcome to the surrogates separately from other IVF-treated patients in the post-partum period. These are pregnancy-related risks which are additional to the risks they take for their own pregnancies—by exposing themselves to surrogate motherhood.

Reproductive Loss

If surrogates have experienced a previous reproductive loss, such as a miscarriage, abortion or neonatal death, the consequences can be significant and long lasting (Bergner et al. 2008). The psychological impact of reproductive loss such as miscarriages is not always considered in clinical care plans (van den Akker 2011) and this is even less likely to be considered in surrogate mothers. It is not only the personal, pre-surrogacy reproductive loss which needs to be recognised and addressed in policy and practice of surrogate arrangements but also any loss they may experience in their surrogate pregnancies, which is currently minimally addressed. Research has shown that not all individuals are well supported within their social networks and some surrogates are known to lack social support (van den Akker 2007a). Culture-specific needs of surrogates across the world experiencing these diverse reproductive losses need to be considered in care plans as research has shown people's needs are different (Haws et al. 2010; Hsu et al. 2002) which should be reflected in the health care they receive. The potential impact of reproductive loss for surrogate mothers may be as substantial as is reported in non-surrogate pregnancies, although no research on embryo 'loss' or feelings of loss of gametes and severe psychological consequences exists (Daugirdaite et al. 2015). Nevertheless, post-traumatic stress (PTS) and post-traumatic stress disorder (PTSD) following miscarriage, abortion, perinatal loss, stillbirth and neonatal death after non-surrogate pregnancies have been reported (Daugirdaite et al. 2015), suggesting surrogate mothers' experiences of reproductive loss or failures may also need to be investigated and supported as the long-term consequences can be devastating and severe.

Attachment and Bonding

Much research has been carried out into maternal-foetal attachments or 'bonding', and the evidence of its importance receives some support in pregnant women carrying a baby which they wished to keep (Cranley 1981; Lerum and LoBiondo-Wood 1989). However, to what extent this is generalisable to surrogate mothers who conceive for the sole purpose of relinquishing the baby is not known. Fischer and Gillman (1991) conducted a study of 21 surrogate mothers. Their main question concerned the relationship between a woman's ability to give up a baby and the degree to which they were attached to the baby. Fisher and Gillman's study was guided by the research relating to surrogates' feelings about pregnancy. As was noted earlier (particularly from Parker and Hanifin's list of motivating factors), a desire to be pregnant and enjoyment of the pregnancy state were common factors. The authors made the underlying assumption that if surrogate mothers can relinquish a child, she is less likely to be attached to the foetus during pregnancy compared to non-surrogate pregnant women. They demonstrated less attachment to the foetus in surrogate compared to non-surrogate pregnant mothers. Fischer and Gillman (1991) interpret this finding as probably related to the use of denial of the importance of developing an attachment to the foetus. The denial would be a conscious intention to distance themselves from attachment.

Muller (1996) reported an association between prenatal attachment and post-natal attachment styles, and as is shown in Chap. 6 on separation and parenting and Chap. 7 on infants, attachment in early infancy is associated with the development of the child as it grows up (Bowlby 1969). Conceptually pre- and post-natal attachments are somewhat different. Prenatal attachment involves a cognitive non-interactional adjustment from someone who needs care to becoming a caregiver (Solomon and George 1996) in preparation for the post-natal period where the demands become a reality. In post-natal attachment nurturing and caregiving behaviours are necessary in reciprocal stimulus—responses dynamic and changing interactions and demands between the mother and infant (Solomon and George 1996). Since prenatal attachments are relatively non-interactional, Laxton-Kane and Slade (2002) suggest these attachments are based upon the mother's own attachment experiences as well as the impact of potentially competing roles such as being cared for to becoming the carer. However, in the surrogate mother's pregnancy, she knows she will not be the carer of the foetus. The intensity (mother's pre-occupation with the foetus) and quality (closeness/distance, tenderness/irritation and positive/negative feelings towards the foetus) of her attachment to the foetus are therefore of some theoretical interest, as these may be minimal or non-existent in surrogate pregnancies.

Attachment between a pregnant woman and her foetus is commonly reported (Cranley 1981). There was some speculation about the existence of a sensitive period when a mother attaches to her foetus which may be triggered by the endocrine changes she experiences as a result of the pregnancy and the movements of the baby (Klaus and Kennell 1982; Minde 1986), although that has not been confirmed. Instead a mother's relationship with her own mother may be a better predictor of prenatal attachment (Wayland and Tate 1993). Research has shown that surrogate mothers can successfully restructure their thinking about the foetus ensuring the 'distance and the involvement required' for relinquishment of a surrogate baby are successful (Snowdon 1994, p. 83). Similar results are reported in other research (Fischer and Gillman 1991; Baslington 2002; van den Akker 2005), and Chesler (1988) goes further and notes that a surrogate mother's reasons for embarking on an arrangement probably has little to do with feelings of attachment. Surprisingly, in Thornton et al.'s (1994) study, where 50 non-surrogate women were asked if they would find it harder to relinquish a genetic or gestational surrogate baby, no differences were reported. However, these participants were not involved in surrogate arrangements. In contrast, van den Akker (2003) found substantial differences in actual surrogate and commissioning mothers, with genetic surrogates mostly claiming they would not find it harder to relinquish a genetic baby, and nearly all gestational surrogates believing they would find that much harder. In the same vein, the commissioning mothers using gestational surrogacy believed, like their surrogate counterparts, that it would be much harder to relinquish a genetically related baby than the commissioning mothers using genetic surrogacy. Similar views were expressed by Blyth's (1994), and Snowdon's (1994) surrogates. There is no information on the effects of increased or decreased prenatal attachment and outcomes for the infant and the mother in surrogate pregnancies, marking yet another gap in psychological research and theory.

In traditional pregnancies prenatal attachment appears to increase as the pregnancy progresses, and this may be related to both awareness of quickening (early movements in the second trimester) and amount of movement, although it is not clear if vigilance or amount of prenatal attachment increases movement detection (Laxton-Kane and Slade 2002). There is no known effect for whether a pregnancy is planned or having had a scan and prenatal attachment, but there is some suggestion that a higher investment in a pregnancy and higher risks possibly cancel each other out (Laxton-Kane and Slade 2002). Parity (Berryman and Windridge 1996) and IVF conception versus natural conception (McMahon et al. 1997) do not influence prenatal attachment, whereas increased levels of social support do seem to affect prenatal attachment. Finally, surrogates attach less to the foetus (Fischer and Gillman 1991) and have less concerns and worries about the foetus during pregnancy (van den Akker 2007b) than commissioning mothers.

According to McGee (1997), 'risks or experiences of bonding are in the order of routine investment risks, manageable through the normal negotiating tools of the market: persuasion, remuneration, contracts and litigation', and surrogacy tends to be 'sealed' by contracts because nobody trusts anybody else (p. 408). Surrogates are trained to see this growing foetus in utero as not theirs, to be detached and to call the baby that of the commissioning parent(s). This is not as easy as it sounds because the foetus is growing and interacting with the mother who is constantly aware of the changes occurring within her body. Post-delivery, this does not diminish as we know from studies of women who gave their babies up for adoption (Boucher et al. 1991). Women are biologically primed to care for and nurture the baby they give birth to and severing that process takes time and emotional effort, regardless of the cognitive preparations which are made in advance of the delivery. Furthermore, no amount of rationalisation can prepare a mother for the tumultuous hormonal changes and their effects upon the birthing mothers' emotional state, although people clearly try.

Nevertheless, studies report the majority of surrogates fare well in the short term with relinquishment of the baby (Blyth 1994; van den

Akker 2003; Ragone 1994; Baslington 2002). Unfortunately, alongside a number of reports of detachment of surrogate mothers' views of their surrogate babies, an equal number of surrogate mothers disclosed severe discomfort following separation from the baby. Intense emotional pain and prolonged crying were reported by some of Ragone's (1994, p. 79) surrogate mothers (particularly those in 'closed' programmes where they did not interact with the intended commissioning couples), and emotional distress and an 'urge' to bond with the baby were reported by some of Ciccarelli (1997, p. 56) and Ciccarelli and Beckman's (2005) surrogate mothers. These reports of distress following relinquishment of the surrogate baby are commonly reported in the psychological and social research literature (Blyth 1994; van den Akker 2007b; Baslington 2002) and show the diversity of responses. The fact that there is no universal good feeling post-delivery in surrogate mothers warrants further research.

According to Baslington's (2002) case study, she proposes a 'maternal-foetal detachment theory' based upon payment. She refers to differences with birth mothers relinquishing babies in adoption, and the long-term emotional effects of relinquishment associated with (particularly closed) adoption (Boucher et al. 1991). Importantly the differences are huge, with the surrogate pregnancy being a chosen pregnancy, not previously existing, and surrogate mothers tend to be slightly older, usually parous and counselling is offered throughout (although not taken up as much as they could). Adoptive mothers on the other hand tend to have unplanned and not chosen pregnancies at a young age and find themselves without sufficient support services. Self-help groups and organisations supporting surrogates in some countries where surrogacy is either legal or tolerated tend to be beneficial as they allow for reference points, 'governing members' attitudes, values and interactions' (Levy 1979, p. 255), and buffer feelings of sadness or regret.

Relinquishment of the Baby

Few surrogates keep the baby. As most surrogates' motivations are altruistic, attention/fulfilment seeking or for financial remuneration, it is not surprising they do not wish to bring up a baby they had not intended to keep when they went into the arrangement. Only in a few cases does this happen. If a surrogate mother uses genetic surrogacy, emotionally one could argue it may be harder to give up a child 'which is half mine' (Snowdon 1994). In gestational surrogacy, this is not a factor. However, in both cases, some bonding as a result of the pregnancy could have developed, and in both cases there are surrogates known to renege on the arrangement (see e.g. genetic surrogacy—Gayle 2015; gestational surrogacy—Nye 2013). There is little or no research which has investigated the possibility of differential bonding and the ability to relinquish a baby. Data from a prospective six-month longitudinal follow-up study of surrogate and commissioning mothers has revealed that few surrogates believe they are bonding with the baby during the pregnancy (van den Akker 2003).

Even following delivery, it is possible that some deep emotional attachment develops, making it difficult for the surrogate mother to relinquish the surrogate baby. Some do indeed look at the infant and feel traumatised by having to give the baby up to the waiting couple, but most do it despite their personal feelings. COTS, the United Kingdom's main surrogate agency, estimate that well over 90% of their surrogate arrangements have reached successful outcomes. Successful conclusions constitute keeping to the arrangement and relinquishing the baby following delivery or the mutually agreed time. However, even if a successful outcome in practice is measured by rates of surrogates relinquishing babies, the post-relinquishment psychological wellbeing of the surrogate also needs to be considered. Women relinquishing a baby for adoption are reported to suffer psychological distress which can last a long time (Baslington 2002). In surrogacy, post relinquishment in the short term is reported to be tainted by feelings of unhappiness (Blyth 1994; Ragone 1994; van den Akker 2003) guilt, regret and loss (Baslington 2002; van den Akker 2007a). Little evidence exists of the long-term effects beyond six months post-delivery, but there are reports of feeling devastated, particularly in genetic surrogacy where contact with the commissioning couple was reported as poor or non-existent (Cotton 1985). However, although the quality of the relationship (or no relationship) with the commissioning parents may play a role in the post-relinquishment psychological health of the surrogate mothers (as in adoption), it is also possible that individual psychological or social factors are relevant.

The successful cognitive restructuring of the act of relinquishing the surrogate baby has been found to be important as shown above. In social terms, it is also possible that the subcultural (van den Akker 2007a), political (Baslington 2002) and media pressures (van den Akker 2016a, b) have an effect on the longer-term psychological health of the surrogate mother. In 1998, Smith described the 'hostile reaction which has been reported in the media, provoked by surrogacy'. She argues this is a direct result of 'society taking for granted the natural bond between mother and child, rejecting anything that may threaten it (p. 188). Incorporating the sociocultural contexts therefore becomes critical in developing an understanding of the processes involved in surrogate motherhood. For example, De Beauvoir's (1953, p. 490) political argument that there is no such thing as 'a maternal instinct' was deemed a revolutionary concept in post-Victorian Britain, although it was not accepted by mainstream citizens. Nevertheless, it found some support in historical research (Badinter 1981). The proposal that a maternal 'instinct' does not exist, coupled with evidence for the social reasons for and approach to parenting and the collective attitude to child sharing in many African countries (Dyer et al. 2008; Dimka and Dein 2013), should, like paid surrogacy, 'break the myth of the maternal instinct' (Roach Anleu 1990, p. 72). This suggests that if society is ready to bust the myth of the maternal instinct, surrogate motherhood may have very different effects on the women relinquishing babies.

Genetic Link

Denial of the importance of the attachment bond (described earlier) with a surrogate baby is likely to be a realistic explanation of ease of relinquishment of a baby, because other research has demonstrated a similar degree of active cognitive restructuring of the importance of (genetic) 'motherhood' in surrogates. For example, Stevens and Dally (1985) reported on a deliberate mental effort made by a surrogate to become more detached from the foetus, and van den Akker found that surrogates changed their views in relation to their perception of the importance of a genetic link according to the type of surrogacy they chose to do (van den Akker 2003, 2007a). In one study, 100% of genetic surrogate mothers reported they did not think a genetic link was important in a child—hence their ability to relinquish a genetically related baby, whereas 72% of gestational surrogate mothers believed a genetic link was important—who knew they could not relinquish a baby genetically related to them. Baslington (2002) refers to this dissonance as a learning process, and these processes are reiterated by Snowdon (1994) who refers to the selective invoking or rejecting of salient biological facts as befits their situation. Ragone (1998, 1999) too reported on the intended mother's need to emphasise the importance of nurturing and social motherhood, whereas the surrogate mother de-emphasises biogenetic ties to the infant as noted in Chap. 5. She reports on similar distancing processes in oocyte donors, to diminish the importance of the biogenetic connection (Ragone 1999, p. 82).

Further research in this area is needed. If future research can demonstrate that intent prior to pregnancy makes little difference to bonding, or maternal-foetal attachment, the practice of surrogacy needs to be reconsidered. This would not necessarily need to constitute a ban on the practice to protect the emotional wellbeing of the surrogate mother, but could involve greater post-natal care of the surrogate mother and perhaps more considerate pre-delivery preparation of the loss following delivery which might have a negative impact upon the surrogate. The fact that proportionally surrogate pregnancies and deliveries are relatively few and far between does not preclude them from having workable and evidence-based pre- and post-natal care plans. Ultimately surrogate pregnancies are becoming increasingly popular and they are brought about via health care and medical professionals. These same medical and health care workers base their non-surrogate pregnancies on evidence-based care plans, indicating this must be reconsidered for surrogate pregnancies since their needs are shown to be different.

Social Support

To date two quantitative studies of surrogate mothers have incorporated the concept of social support in their research. Fischer and Gillman (1991) argued that the measurement of social support would not only provide information of a surrogate's general pattern of relatedness, but also as a measure of her patterns of attachment. van den Akker (2003, 2007a) measured social support more as a measure of general health and wellbeing in relation to the development of psychopathology. Since giving birth and relinquishing a baby are likely to be major (even stressful) events, less social support should be related to more psychopathology post relinquishment of the baby, and the reverse would also be true. Fisher and Gillman found that although numbers and types of problems experienced during the six months prior to psychological testing were similar in the surrogate and non-surrogate mothers, surrogates sought less social support than non-surrogate pregnant mothers. Particularly notable was a significantly different absence of family support from parents, spouse and children within the surrogate group. van den Akker (2003) revealed surrogate mothers were often not well supported by their parents or partner although many did not have a stable partner. Friends tended to be more supportive than family. This absence of family support may be particularly salient in genetic surrogates as they are in effect relinquishing a baby which is genetically related to the extended family. Grandparents will miss out on grandparenting the surrogate child, as do aunts, uncles and cousins. The lack of opportunity to take a new child into the support system of the extended family is less likely to be an issue for the extended family in gestational surrogate motherhood, as the baby is not genetically related to the surrogate mother's family. To date, research on the implications of assisted conception upon the wider society (van den Akker 2016) and of surrogate motherhood on the extended family is non-existent, but research on the needs of adult donor conceived individuals, their half siblings and their donors has shown these are visceral in many (van den Akker et al. 2015). Furthermore, based upon the lack of parental support given to surrogate mothers, negative intergenerational effects are likely to exist.

In van den Akker's (2003) study, surrogates reported receiving some support from family, and some from agencies, though little support was received from clinics, GPs or counsellors. One surrogate mother had experienced relationship problems with her partner during pregnancy and post-delivery, but none reported sexual problems. Although in this study all surrogates had informed their family about becoming a surrogate mother, two surrogate mothers had not disclosed this to their friends and colleagues. Those who had not told reported feeling bad about lying and being secretive; one revealed she had told people she experienced a neonatal death. They attributed their dishonesty to a general ignorance in those around them about surrogacy and feared stigmatising comments.

Marital relationships were not different in the non-surrogate and surrogate pregnant mothers in Fischer and Gillman's (1991) study, although, as was found by Blyth (1994) and van den Akker (2003), surrogates were less likely to be married or in stable relationships, suggesting these results may be confounded and should be interpreted with caution. Further prospective longitudinal research following surrogates through the six months post-delivery and relinquishment of the baby comes from van den Akker (2005). In this study, post-natal depression at six days, six weeks and six months post-delivery was present in some, albeit rarely, and anxiety did not increase significantly over time in surrogate mothers.

In studies of cross-border surrogacy such as India, stigma and being ostracised by family, friends and the community have been reported (Karandikar et al. 2014). Some surrogate mothers even had to move to another village to escape the criticism from the community; others were removed from the local church. Lying to family and friends and staying away for the duration of the pregnancy in the clinic/hostel was the only way to get around the stigma of surrogate motherhood (Pande 2010; Twine 2011). However, this has consequences for the surrogate mother's emotional wellbeing and potentially that of her own biological children. The women themselves mentioned missing their homes and children. The existing children of surrogate mothers have not yet been sufficiently studied (Jadva and Imrie 2014), posing a grave omission in the research literature. Social support for some surrogates in some countries is therefore not always offered by those close to them.

Surrogate Motherhood Within the Context of the Family

Interestingly, surrogacy exemplifies the importance of kinship and family, yet it is exactly family and family relationships that are fragmented for the surrogate who is trying to 'patch up' the fragmentation of the commissioning parent(s). The surrogate mother's own children are understudied, but it is recognised that there could be implications for them, particularly if the mother is hospitalised or living away from home for some of the pregnancy and post-delivery. The ASRM Practice Committee Guideline on Gestational Carriers (2012) recommends that the psychosocial evaluation and counselling of the surrogate mother should also consider the impact of the surrogate's pregnancy on 'family and community dynamics' (p. 1304). Research which has asked surrogates what the impact of surrogacy would have on their own children has been largely problem free (van den Akker 2003) or positive (Jadva et al. 2003). However there is recognition for a need for new research on the impact of surrogacy on surrogates' own children (van den Akker 2007a, 2012; Horowitz et al. 2010). Little research of a surrogate's own children has been published (Jadva and Imrie 2014). Most children were positive about their mother's surrogacy but some were neutral or ambivalent. However, more studies need to be carried out on the children of surrogates because data is already emerging on the stigma associated with surrogacy in India, and the upheaval experienced by the surrogate's own children when she is removed from her home for the duration of the pregnancy (Pande 2010). Furthermore, the extended family such as the surrogate's own brothers, sisters and parents also need to be studied because in genetic surrogacy they all lose a genetically related member of their particular kinship. Grandparents, in particular, may mourn the loss of a grandchild, as may half siblings in genetic surrogacy.

According to Delaney (1986), paternity, not maternity, has been conceptually given the status of creation and engendering life; maternity merely nurtures and gives birth to that life. This has been used to define or reduce surrogate motherhood to a nurturing role, and they are thereby seen to be doing something good (Collier et al. 1982), not contradictory in motherhood terms. The nurture/culture dichotomy evaporates with the reconstructed interpretation of surrogate motherhood, that of providing the biological nurturing role—not the important social motherhood role which is reserved for the commissioning mother of the surrogate baby. Thus, between the two women a series of conceptual restructuring of important cognitions develops, leaving both with the feeling that true motherhood is not represented by the nurturing surrogate, but the social commissioning mother or parent. The problematic biogenetic relationship between the surrogate mother and the baby and the lack of this important biogenetic relationship between the commissioning mother and the surrogate baby are solved with these new meanings (Ragone 1996). Surrogate mothers reconstruct their own role in fragmenting kinship through becoming a surrogate by highlighting only that which is consonant with traditional family ideology. They de-emphasise that which contradicts aspects of family ideology, and re-emphasise other aspects that buffer their status as kinship valuers (e.g. by emphasising a desire to help infertile couples become a much-valued family). This disperses the potential conflict they could experience between their own traditional role as mothers and their untraditional role as surrogates.

The future of reproductive health care services should reflect the specific, lifetime and shifting needs of the populations they serve, including future generations resulting from these innovations. Gamete and embryo donors have no legal parenthood status and are therefore not named on birth certificates and are not financially liable. They also have no legal obligation to the child, or have rights on its upbringing (HFEA 2016). In contrast surrogate mothers are the legal mothers of surrogate children born in the United Kingdom, because birth motherhood determines motherhood, not genetics as is the case in some other jurisdictions (see Chap. 9).

Summary

In short, research efforts to date have demonstrated that surrogate mothers have various reasons for becoming surrogates. No differences in those who get primarily involved in surrogacy arrangements for the benefit of money or for altruistic reasons are apparent. Furthermore, the experience of surrogacy is unique and should be interpreted differently from pregnancy in women who are non-surrogates. Personality profiles are not significantly different from the general population and psychological functioning prior to, during and following delivery is not significantly compromised in surrogate mothers. The most negative aspects of the surrogacy experience include some difficulty in coping with the relinquishment of the baby for a proportion of surrogates, and a larger proportion reporting missing the contact surrogates enjoyed with the commissioning couples. It is likely that the attachment formed with the commissioning couple was produced as a deliberate shift from an attachment to the baby, to justify their relinquishment. Thus, the feeling of loss of and longing for the baby gets translated into a feeling of loss of and longing for the closeness of the distancing commissioning couple. The long-term well-being of surrogate mothers needs further research and pre-arrangement counselling should prepare surrogates for the possibility of experiencing this kind of grief. The evidence thus far therefore suggests that overall surrogacy works from the point of view of the surrogate mothers. However, major reviews of surrogate motherhood arrangements, 10 years apart, note that much of the research suffers from methodological limitations (van den Akker 2006, 2007b; Söderström-Anttila et al. 2016). The next chapter will consider the research into prospective parents commissioning surrogate babies.

References

Andrews, A. (1982). Stork market, supra note 7; K. M. Brophy. A surrogate mother contract to bear a child. *University of Louisville Journal of Family Law, 20*, 263–291. OTA, Infertility, supra note 1.

ASRM. (2012). Practice Committee of the Society for Assisted Reproductive Technology. Recommendations for practices utilizing gestational carriers: An ASRM practice committee guideline. *Fertility and Sterility, 97*, 301–308.

Badinter, E. (1981). *The myth of motherhood: An historical view of the maternal instinct*. London: Macmillan.

Baslington, H. (2002). The social organisation of surrogacy: Relinquishing a baby and the role of payment in the psychological detachment process. *Journal of Health Psychology, 7*, 58.

BBC 4. (2013). House of surrogates. http://www.bbc.co.uk/programmes/b03c591s. Accessed 26 Jan 2016.

Bergner, A., Beyer, R., Burghard, F., Klapp, F., & Rauchfuss, M. (2008). Pregnancy after early pregnancy loss: A prospective study of anxiety and depressive symptomatology and coping. *Journal of Psychosomatic Obstetrics and Gynecology, 29*(2), 105–113.

Berryman, J. C., & Windridge, K. C. (1996). Pregnancy after 35 and attachment to the foetus. *Journal of Reproductive and Infant Psychology, 14*, 133–143.

BioNews. (2005, January 31). Surrogate dies in childbirth. *BioNews,* (293). http://www.bionews.org.uk/page_12239.asp. Accessed 5 Oct 2016.

Blyth, E. (1994). "I wanted to be interesting, I wanted to be able to say 'I've done something with my life'": Interviews with surrogate mothers in Britain. *Journal of Reproductive and Infant Psychology, 12,* 189–198.

Blyth, E., Crawshaw, M., Frith, L., & van den Akker, O. (2017). The modern family and the future? Gamete donors' motivations for, expectations and experiences of, registration with UK donor link. *Human Fertility,* 1–11. doi: 10.1080/14647273.2017.1292005.

Boucher, P., Lambert, L., & Triseliotis, J. (1991). *Parting with a child for adoption: The mother's perspective.* London: British Agencies for Adoption and Fostering (BAAF).

Bowlby, J. (1969). *Attachment, Vol. 1 of Attachment and loss.* New York: Basic Books.

Brinsden, P. R. (2003). Gestational surrogacy. *Human Reproduction Update, 9,* 483–491.

British Infertility Counselling Association. http://bica.net/about. Accessed 29 Apr 2016.

Bromfield, N., & Smith-Rotabi, K. (2014). Global surrogacy, exploitation, human rights and international private law: A pragmatic stance and policy recommendations. *Global Social Welfare, 1*(3), 123–135.

Bruno, C., Dudkiewicz-Sibony, C., Berthaut, I., Weil, E., Brunet, L., Fortier, C., et al. (2016). Survey of 243 ART patients having made a final disposition decision about their surplus cryopreserved embryos: The crucial role of symbolic embryo representation. *Human Reproduction, 31*(7), 1508–1514.

Burrell, C., & Edozien, L. (2014). Surrogacy in modern obstetric practice. *Seminars in Foetal and Neonatal Medicine, 19,* 272–278.

Charo, R. A. (1988). Legislative approaches to surrogate motherhood. *Law, Medicine and Health Care, 16*(1–2), 96–112.

Chesler, P. (1988). *Sacred Bond: Motherhood under siege.* London: Virago.

Ciccarelli, J. (1997). *The surrogate mother: A post-birth follow-up study.* Los Angeles: California School of Professional Psychology.

Ciccarelli, J., & Beckman, L. (2005). Navigating rough waters: An overview of psychological aspects of surrogacy. *Journal of Social Issues, 61*(1), 21–43.

Clark, E., & Silver, R. (2011). Long term maternal morbidity associated with repeat caesarean delivery. *American Journal of Obstetrics and Gynecology, 205,* S2–S10.

Collier, J., Rosaldo, M., & Yanagisako, S. (1982). Is there a family? In B. Thorne & M. Yalom (Eds.), *Rethinking the family.* New York: Longman.

Corea, G. (1985). *The mother machine: Reproductive technologies from artificial insemination to artificial wombs*. New York: Harper & Row Publishers.

Corson, S. L., Kelly, M., Braverman, A. M., et al. (1998). Gestational carrier pregnancy. *Fertility and Sterility, 69*, 670–674.

Cotton, K. (1985). *Baby Cotton: For love or money*. London: Dorling Kindersley. London.

Cotton, K. (1992). *Second time around. The full story of my second surrogate pregnancy*. Printed by Dornoch Press Limited. London.

Cranley, M. (1981). Development of a tool for the measurement of maternal attachment during pregnancy. *Nursing Research, 30*(5), 281–284.

Crawshaw, M., Blyth, E., & van den Akker, O. (2012). The changing profile of surrogacy in the UK – Implications for policy and practice. *Journal of Social Welfare and Family Law, 34*, 1–11.

Crawshaw, M., Frith, L., van den Akker, O., & Blyth, E. (2016). Voluntary DNA-based information exchange and contact services following donor conception: An analysis of service users' needs. *New Genetics and Society, 35*(4), 372–392.

Daugirdaite, V., van den Akker, O.. & Purewal, S. (2015). Posttraumatic stress and posttraumatic stress disorder after termination of pregnancy and reproductive loss: A systematic review. *Journal of Pregnancy, 2015*, 646345, 14 pages.

De Beauvoir, S. (1953). *The second sex*. London: Jonathan Cape.

De Lacey, S. (2013). Decision making about frozen supernumerary human embryos. *Human Fertility, 16*, 31–34.

Delaney, C. (1986). The meaning of paternity and the virgin birth debate. *Man, New Series, 21*(3), 494–513.

Dimka, R. A., & Dein, S. L. (2013). The work of a woman is to give birth to children: Cultural constructions of infertility in Nigeria. *African Journal of Reproductive Health, 17*(2), 102–117.

Duffy, D., Nulsen, J., Maier, D., Engman, L., Schmidt, D., & Benadiva, C. (2005). Obstetrical complications in gestational carrier pregnancies. *Fertility and Sterility, 83*, 749–754.

Dyer, S., Mokoena, N., Maritz, J., & van der Spuy, Z. (2008). Motives for parenthood among couples attending a level 3 infertility clinic in the public health sector in South Africa. *Human Reproduction, 23*(2), 352–357.

Egliston, K., McMahon, C., & Austin, M.-P. (2007). Stress in pregnancy and infant HPA axis function: Conceptual and methodological issues relating to the use of salivary cortisol as an outcome measure. *Psychoneuroendocrinology, 32*(1), 1–13.

Fischer, S., & Gillman, I. (1991). Surrogate motherhood: Attachment, attitudes and social support. *Psychiatry, 54*, 13–20.

Franks, D. D. (1981). Psychiatric evaluation of women in a surrogate mother program. *American Journal of Psychiatry, 138*(10), 1378–1379.

Gayle, D. (2015, May 6). High court orders surrogate mother to hand baby to gay couple. *The Guardian.* https://www.theguardian.com/law/2015/may/06/high-court-orders-surrogate-mother-baby-gay-couple. Accessed 5 Oct 2016.

Hanifin, H. (1987). *Surrogate parenting: Reassessing human bonding.* Paper presented at the annual meeting of the American Psychological Association.

Haws, R. A., Mashasi, I., Mrisho, M., Schellenberg, J. A., Darmstadt, G. L., & Winch, P. J. (2010). "These are not good things for other people to know" @ how rural Tanzanian women's experiences of pregnancy loss and early neonatal death may impact survey data quality. *Social Science & Medicine, 71*, 1764–1772.

Henaghan, M. (2013). International surrogacy trends: How family law is coping. *Australian Journal of Adoption, 7*(3), 1–24. Retrieved from http://www.nla.gov.au/openpublish/index.php/aja/article/view/3188.

HFEA. (2016). The human fertilisation and embryology authority. http://www.hfea.gov.uk/donor-conception-births.html. Accessed 3 May 2016.

Hinson, D., & McBrien, M. (2011). Surrogacy across America: Both the law and the practice. *Family Advocate, 34*(2), 32–36.

Horowitz, J. E., Galst, J. P., & Elster, N. (2010). *Ethical dilemmas in fertility counseling.* Washington, DC: American Psychological Association.

Horsey, K. (2015). Surrogacy in the UK: Myth busting and reform. Report of the Surrogacy UK. Working Group on Surrogacy Law Reform (Surrogacy UK, November).

Hsu, M., Tseng, Y., & Kuo, L. (2002). Transforming loss: Taiwanese women's adaptation to stillbirth. *Journal of Advanced Nursing, 40*, 387–395.

Jadva, V., & Imrie, S. (2014). Children of surrogate mothers: Psychological well-being, family relationships, and experiences of surrogacy. *Human Reproduction, 29*, 90–96.

Jadva, V., Murray, C., Lycett, E., MacCallum, F., & Golombok, S. (2003). Surrogacy: The experiences of surrogate mothers. *Human Reproduction, 18*(10), 2196–2204.

Karandikar, S., Gezinski, L., Carter, J., & And Kaloga, M. (2014). Economic necessity or noble cause? A qualitative study exploring motivations for gestational surrogacy in Gujarat, India. *Affilia, 29*(2), 224–236.

Klaus, M., & Kennell, J. (1982). *Parent-infant bonding* (2nd ed.). St Louis: C.V. Mosby Co.

Laxton-Kane, M., & Slade, P. (2002). The role of maternal prenatal attachment in a woman's experience of pregnancy and implications for the process of care. *Journal of Reproductive and Infant Psychology, 20*(4), 253–266.

Lerum, C. W., & LoBiondo-Wood, G. (1989). The relationship of maternal age, quickening and physical symptoms of pregnancy to the development of maternal–fetal attachment. *Birth, 16,* 13–17.

Levy, L. H. (1979). Processes and activities in groups. In M. A. Lieberman, L. D. Borman, & Associates (Ed.), *Self-help groups for coping with crisis. Origins, members processes and impact* (pp. 234–271). San Francisco: Jossey-Bass.

Mansour, R., Ishihara, O., Adamson, G. F., Dyer, S., de Mouzon, J., Nygren, K. G., Sullivan, E., & Zegers-Hochschild, F. (2014). International Committee for Monitoring Assisted Reproductive Technologies world report: Assisted reproductive technology 2006. *Human Reproduction, 29,* 1536–1551.

Masten, A. S., Best, K. M., & Garmezy, N. (1990). Resilience and development: Contributions from the study of children who overcome adversity. *Development and Psychopathology, 2,* 425–444.

McGee, G. (1997). Trials and tribulations of surrogacy legislating legislation. *Human Reproduction, 12*(3), 407–408.

McKauley, K., Elsom, S., Muir-Cochrane, E., & Lyneman, J. (2011). Midwives and assessment of perinatal mental health. *Journal of Psychiatric and Mental Health Nursing, 18,* 786–795.

McMahon, C., Ungerer, J., Beaurepaire, J., Tennant, C., & Saunders, D. (1997). Anxiety during pregnancy and foetal attachment after in-vitro-fertilization conception. *Human Reproduction, 12,* 176–182.

Minde, K. (1986). Bonding and attachment: Its relevance for the present day clinician. *Developmental Medicine and Child Neurology, 28*(6), 803–806.

Muller, M. E. (1996). Prenatal and postnatal attachment: A modest correlation. *Journal of Obstetric, Gynaecologic and Neonatal Nursing, 25,* 161–166.

Nye, J. (2013). Surrogate mother fled across country to give birth and save baby after parents wanted child aborted when ultrasound revealed series of disabilities. Mail online 5 March. http://www.dailymail.co.uk/news/article-2288237/Woman-paid-22-000-surrogate-mother-refuses-parents-wishes-abortion-ultrasound-reveals-series-disabilities-flees-country-save-baby.html. Accessed 5 Oct 2016.

Ombelet, W., De Sutter, P., Van der Elst, J., & Martens, G. (2005). Multiple gestation and infertility treatment: Registration, reflection and reaction – The Belgian project. *Human Reproduction Update, 11,* 3–14.

Pande, A. (2009). 'Not an angel' not 'a whore': surrogates as 'dirty workers' in India. *Indian Journal of Gender Studies, 16,* 1410173.

Pande, A. (2010). Commercial surrogacy in India: Manufacturing a perfect mother-worker. *Signs, 35,* 969–992.

Parker, P. J. (1983). Motivation of surrogate mothers: Initial findings. *American Journal of Psychiatry, 140*(1), 117–118.

Parker, P. J. (1984). Surrogate motherhood, psychiatric screening and informed consent, baby selling, and public policy. *Bulletin of the American Academy of Psychiatry Law, 12*(1), 21–39.

Parkinson, J., Tran, C., Tan, T., Nelson, J., Batzofin, J., & Sarafini, P. (1999). Perinatal outcome after in-vitro fertilization-surrogacy. *Human Reproduction, 14*(3), 671–676.

Pizitz, T., McCullaugh, J., & Rabin, A. (2013). Do women who choose to become surrogate mothers have different psychological profiles compared to a normative female sample? *Women and Birth, 26,* e15–e20.

Ragone, H. (1994). *Surrogate motherhood: Conception in the heart.* Boulder: Westview Press.

Ragone, H. (1996). Chasing the blood tie: Surrogate mothers, adoptive mothers and fathers. *American Ethnologist, 23*(2), 352–365.

Ragone, H. (1998). Incontestable motivations. In S. Franklin & H. Ragone (Eds.), *Reproducing reproduction: Kinship, power and technological innovation.* Philadelphia: University of Pennsylvania Press.

Ragone, H. (1999). The gift of life. Surrogate motherhood, gamete donation, and construction of altruism. In L. Layne (Ed.), *Transformative motherhood: On giving and getting in a consumer culture* (pp. 65–88). New York: New York University Press.

Reame, N. (1989). Maternal adaptation and psychologic responses to a surrogate pregnancy. *Journal of Psychosomatic Obstetrics and Gynecology, 10*(suppl), 86.

Roach Anleu, S. L. (1990). Reinforcing gender norms: Commercial and altruistic surrogacy. *Acta Sociologica, 33*(1), 63–74.

Savasi, V. M., Mandai, L., Laoreti, A., & Cetin, I. (2016). Maternal and foetal outcomes in oocyte donation pregnancies. *Human Reproduction Update, 22*(5), 620–633.

Silver, R. (2012). Implications of the first caesarean: Perinatal and future reproductive health and subsequent caesareans, placentation issues, uterine rupture risk, morbidity and mortality. *Seminars in Perinatology, 36,* 315–323.

Snowdon, C. (1994). What makes a mother? Interviews with women involved in egg donation and surrogacy. *Birth, 21*(2), 77–84.

Söderström-Anttila, V., Blomqvist, T., Foudila, T., Hippelainen, M., et al. (2002). Experience of in vitro fertilization surrogacy in Finland. *Acta Obstetrica Gynecologia Scandinavica, 81,* 747–752.

Söderström-Anttila, V., Wennerholm, U., Loft, A., Pinborg, A., Aittomak, K., Romundstad, K., & Bergh, C. (2016). Surrogacy: Outcomes for surrogate mothers, children and the resulting families – A systematic review. *Human Reproduction Update, 22*(2), 260–276.

Solomon, J., & George, C. (1996). Defining the caregiving system: Towards a theory of caregiving. *Infant Mental Health Journal, 17*, 183–197.

Stevens, K., & Dally, K. (1985). *Surrogate mothers: One woman's story*. London: Century.

Thornton, J., McNamara, H., & Montague, I. (1994). Would you rather be a 'birth' or a 'genetic' mother? If so, how much? *Journal of Medical Ethics, 20*, 87–92.

Twine, F. (2011). *Outsourcing the womb: Race, class and gestational surrogacy in a global market*. New York: Routledge.

van den Akker, O. B. A. (2003). Genetic and gestational surrogate mothers' experience of surrogacy. *Journal of Reproductive and Infant Psychology, 21*(2), 145–161.

van den Akker, O. B. A. (2005). A longitudinal pre pregnancy to post-delivery comparison of genetic and gestational surrogate and intended mothers: Confidence and genealogy. *Journal of Psychosomatic Obstetrics and Gynecology, 26*(4), 277–284.

van den Akker, O. B. A. (2006). A review of gamete donor family constructs: Current research and future directions. *Human Reproduction Update, 12*(2), 91–101.

van den Akker, O. B. A. (2007a). Psychosocial traits and state characteristics, social support and attitudes to the surrogate pregnancy and baby. *Human Reproduction, 22*, 2287–2295.

van den Akker, O. B. A. (2007b). Psychosocial aspects of surrogate motherhood. *Human Reproduction Update, 13*(1), 53–62.

van den Akker, O. B. A. (2011). The psychological and social consequences of miscarriage. *Expert Review of Obstetrics and Gynecology, 6*(3), 295–304.

van den Akker, O. B. A. (2012). *Reproductive health psychology*. Wiley-Blackwell. ISBN-13: 978-0470683385.

van den Akker, O. B. A. (2013). For your eyes only: Bio-behavioural and psychosocial research objectives. *Human Fertility, 16*(1), 89–93.

van den Akker, O. B. A. (2015). Chapter: Emotional and psychosocial risk associated with fertility treatment. In R. Mathur (Ed.), *Reducing risk in fertility treatment*. London: Springer Science + Media.

van den Akker, O. (2016). Reproductive health matters. *The Psychologist, 29*(1), 2–5.

van den Akker, O. B. A., Crawshaw, M. C., Blyth, E. D., & Frith, L. J. (2015). Expectations and experiences of gamete donors and donor-conceived adults searching for genetic relatives using DNA linking through a voluntary register. *Human Reproduction, 30*(1), 111–121.

van den Akker, O., Camara, I., & Hunt, B. (2016a). Together … for only a moment' British media construction of altruistic non-commercial surrogate motherhood. *Journal of Reproductive and Infant Psychology, 34*(3), 271–281.

van den Akker, O., Fronek, P., Blyth, E., & Frith, L. (2016b). 'This neo-natal ménage à trois' British media framing of transnational surrogacy. *Journal of Reproductive and Infant Psychology, 34*(1), 15–27.

Wayland, J., & Tate, S. (1993). Maternal-foetal attachment and perceived relationships with important others in adolescents. *Birth, 20*, 198–203.

5

Commissioning Parents

Individuals commissioning a surrogate baby embark upon a process of unusual reproduction involving at its most extreme IVF or donor conception and the surrogate gestation and delivery. Universally normative beliefs about reproduction remain conservative, and many traditional and non-traditional families using surrogacy emphasise the importance of genetic or biological kinship. Some are reported to resist acknowledging this fundamental difference from traditional conception. Psychological research contextualising third-party-assisted reproduction generally fails to predict how people cope with the lifestyle changes bringing about the need for some third-party conception and third-party-assisted reproduction itself. This chapter covers the policy context and legal frameworks conflicting with the reality of some current practices. Issues of kinship and effects upon the extended family are also understudied.

There is a dearth of information about intended couples or individuals commissioning a surrogate baby. Becoming a parent has long been recognised as one of the most important milestones in life (Heinicke 1995) leading to self-enhancement at community and generational levels (Brodzinsky 1997). It is therefore unsurprising that individuals who cannot have a family experience substantial psychological distress (van den Akker 2012), and depressed women are less likely to seek treatment for

© The Author(s) 2017
Olga B.A. van den Akker, *Surrogate Motherhood Families*,
DOI 10.1007/978-3-319-60453-4_5

their infertility (Crawford et al. 2017). Commissioning mothers are usually women who cannot carry a pregnancy to term for medical reasons or because they have become medically infertile as a result of their age or treatment for disease. Commissioning fathers are men who are in a relationship with a woman, who cannot carry a pregnancy to term, or with another man, or single men. This means that surrogate motherhood is used to build a family for social or medical reasons; in other words, not all commissioning individuals or couples are infertile—some are involuntarily childless for non-medical reasons. However, what all individuals using surrogacy have in common is that genetic and gestational/biological relationships between the commissioning 'parent(s)' and their children are ambiguous. This raises questions about definitions of parents, the family and 'ownership' of surrogate-born children, particularly if genetic surrogacy or donor gametes/embryos in gestational surrogacy are used.

Becoming a parent through surrogate motherhood places these prospective parents between those opting for infertility treatment and those opting for adoption. They have a potential genetic link with a child which makes them different from adoptive parents, but lack the gestation, which they have in common with adopters of existing children. Conversely, unlike IVF mothers, commissioning mothers do not experience the gestation which IVF parents have, but like IVF parents, they can have a potential genetic link with the child. The commonality with both adoption and IVF is that the routes to parenthood are time consuming, require emotional adjustments and can be expensive. However, commissioning a surrogate baby uniquely places the prospective commissioning parents somewhere in between these other routes to parenthood (van den Akker 2001a, b) although it remains the costliest and therefore an option available only to those who can financially afford it. Commissioning a surrogate baby therefore precludes many others who desperately want a family from achieving this important life goal, unless a close friend or relative is willing to be their surrogate mother.

For years academics have discussed the need for redevelopment of 'the family' in surrogate families (Snowden 1994; van den Akker 2001a), and argued that individuals building families using third-party conception should accept difference rather than shoehorn a non-traditional family into a pseudo-traditional framework (Smolin 2016; van den Akker

2016). It may be that there is so much resistance to acknowledging 'difference' and so many attempts to hide the unusual reproduction where the 'unfamiliar' is introduced into new generations of families, because many of the infertile couples opting for surrogacy have already been devastated by their inability to carry a pregnancy to term. This resistance to divulge difference is not evident in same-sex and solo parents, possibly because their lack of an opposite-sex partner makes that impractical and futile. Non-heterosexual commissioning parents therefore do not necessarily have medical reasons to use surrogacy as a means to (partly) genetic parenthood. Single men have widely different and varied reasons for using surrogacy. For example, drawing on a famous case, the well-known young heterosexual single man Cristiano Ronaldo who commissioned a surrogate baby in 2010 apparently did so to be a young parent. In contrast, a less well known but also very wealthy young Japanese man Mitsutoki Shigeta, aged 24, commissioned 16 surrogate children over a two-year period with no end in sight to his desire to have children. Apparently he wanted 10–15 babies a year until his death for no other reason than he wanted them (Rawlinson 2014). However, wanting to have a surrogate child just because you can is not exclusive to single men. In principle, any one individual who would like to reproduce but cannot do so for social or medical reasons could commission a surrogate baby, although, as is shown in this and other chapters in this book, in practice this is not the case.

Reasons for Commissioning a Surrogate Baby

Medical Factors

The majority of factors responsible for failed conceptions and failed pregnancies are due to embryonic or foetal factors and another proportion to maternal or multifactorial factors; see for example Fig. 5.1. Where embryonic factors are the main cause, IVF could be used to investigate the embryo's development and implantation since implantation rates following the embryo's transfer to the woman's uterus are improving (Margalioth et al. 2006). According to Anderson et al. (2005)'s paper

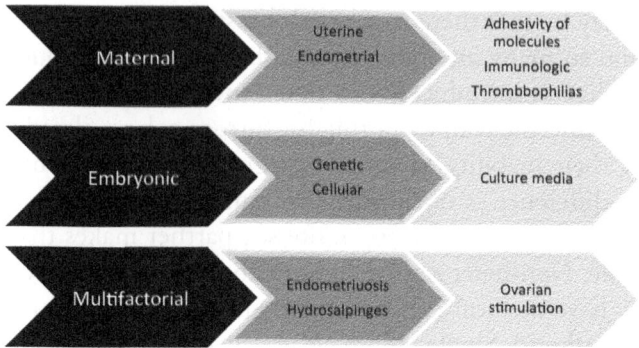

Fig. 5.1 Showing the main aetiologies for repeated implantation failure (Adapted from Margalioth et al. 2006)

on 2001 European ART data, close to 29% of 235,000 treated cycles resulted in a pregnancy. There are differences across different clinics with some achieving better results than others, but despite such differences in techniques and success rates, there are couples for whom the technique does not work and who continue to fail to be able to implant an embryo or fail to sustain a pregnancy.

When embryonic factors such as chromosomal abnormalities are the main cause, prenatal genetic diagnosis or screening can be offered (PGD or PGS). Here chromosomally normal embryos are selected for embryo transfer, and this has led to improved outcomes such as 43% pregnancy rates of and 32% delivery rates in young women (Taranissi et al. 2005), although there is some doubt on the evidence that women with multiple failures to implant benefit greatly from PGD (Caglar et al. 2005).

Maternal Factors

Another, considerably less frequently adapted strategy for treatment, particularly in the case of multifactorial aetiologies of failure to conceive or maintain a pregnancy, includes the personal or psychological state of the individual. There is some suggestion that psychological factors can cause infertility, because of the stress associated with the diagnostic or treatment process. However, the evidence is not yet convincing, with, for example, some research suggesting psychological interventions do not improve pregnancy rates (Boivin 2003), others reporting they do (de Liz and Strauss 2005).

Reproductive Failure and Loss

An alternative to IVF with a couple's own gametes is using IVF with donor gametes and/or embryo transfer with donor embryos could be used. However, if the cause of the failed implantations or failed pregnancies is maternal factors, little can be done to reverse this. In these cases, using another woman to carry the pregnancy is indicated. Similarly, if a woman has a known defect or absence of her uterus through, for example, treatment for disease such as cancer or a hysterectomy, then surrogacy is also a viable medical alternative. Research has shown that many people go through repeat cycles of fertility treatment, some of those give up and many others persevere (Jayakrishnan 2012). A substantial proportion will never succeed in a successful implantation or pregnancy through to a live birth.

The impact of failed pregnancies cannot be underestimated. Miscarriages alone occur in approximately a third of pregnant women (Corbet-Owen and Kruger 2001), is devastating to most of them and these effects can be lasting (Bergner et al. 2008). In a review of the psychological impact and social context of miscarriage (van den Akker 2011) the immediate and long-term recognition of the psychological effects of these losses are not always considered in the care of those experiencing the loss. These need to be recognised and addressed in policy and practice, including in patients experiencing loss of a surrogate pregnancy, loss of an embryo or loss of gametes. In addition to recognising the effects of the losses, there is a need to provide support to women and men (Beutel et al. 1996) as not all are well supported within their social networks (Rosenfeld 1991). Finally, there is a need to address the culture-specific needs of the people experiencing these diverse reproductive losses (Haws et al. 2010; Hsu et al. 2002), as the health care needs may be different for people of different cultures for the same losses.

In a recent systematic review of reproductive loss, PTS and PTSD, Daugirdaite et al. (2015) reported the effects of all possible reproductive losses on the psychological health of the people involved. A total of 48 papers were included in the systematic review covering miscarriage, abortion, perinatal loss, stillbirth and neonatal death. No research on embryo 'loss', or feelings of loss of gametes, was found. Across the different types of loss, women reported more PTS and PTSD following reproductive loss than men. Women with advanced pregnancies, a history of previous trauma, previous mental health problems and adverse psychosocial profiles were also at higher risk of developing PTS and PTSD following reproductive loss. The focus of much research and practice is on 'fixing' and creating and sustaining a pregnancy. These reviews indicate that the substantial losses experienced across reproductive failures are insufficiently studied and insufficiently supported, as the long-term consequences can be devastating and severe.

A case study presented by Raziel et al. (2000) showed the successful use of surrogacy in a woman who had previously had 24 abortions covering a period of just over a decade. This case study presented a woman in whom maternal factors such as systemic or immunologic causes were likely to be responsible for the multiple abortions, and not factors associated with the embryo or foetus. The woman had all her pregnancies with the same partner. Her hysterosalpingography (radiologic investigation of the shape of the uterus and fallopian tubes) and hysteroscopy (endoscopic exploration of uterus via the cervix), and hormonal analysis were normal, and she did not have any autoimmune diseases. Karyotypes (the number and appearance of chromosomes in cell nuclei) in the mother, father and foetus(es) were normal. After several treatments to sustain future pregnancies (e.g. low-dose aspirin, oral dydrogesterone, heparin), she was introduced to surrogacy which resulted in a successful birth of a baby. However, nothing was reported on the woman's long-term experience of the multiple losses prior to the successful surrogate birth, and nothing about her ability to parent the surrogate-born child. The responsibilities of professionals tend to stop once a successful pregnancy or delivery is achieved. No one takes responsibility for the longer-term trajectory, and care plans for these populations which include psychosocial care are urgently needed (van den Akker 2013; Payne and van den Akker 2016).

Social Factors

Commissioning a surrogate pregnancy for non-medical or social reasons is increasingly common. Researchers have debated the ethics of using a surrogate pregnancy for social reasons (e.g. Perla 2001). In Perla's article, examples of older women are used to question whether they are the best patients to receive IVF to conceive a baby, because, according to nature, female fertility declines even before the age of 35, and risks to the mother (including risks of miscarriage) and foetus (including, e.g., chromosomal abnormalities) increase beyond that age. What is of considerably more concern is the fact that many people do not receive fertility treatment on the NHS, particularly not surrogacy. This means building a family using surrogacy is available only to those who can afford it, a point integral to the ethics of third-party conception and is explored further in Chap. 9. Surrogacy, like any other third-party conception, is an option and a treatment for generally healthy people who are unable to build a family for social reasons (single and same-sex couples are included in these), and is therefore not only an option for people with medical conditions. It involves a personal choice to parent a child against 'social' odds. Gestational surrogacy which involves oocyte retrieval from the commissioning mother also involves a potential risk to her, including a risk of medical complications.

Commissioning a Genetically Related or Unrelated Baby

Some commissioning parents are genetically related to the surrogate baby; others are not or are only partially genetically related to the surrogate baby as shown in Fig. 5.2. The final example is likely to be a donated embryo using IVF separately; donated gametes can also be transplanted into a surrogate using embryo transfer.

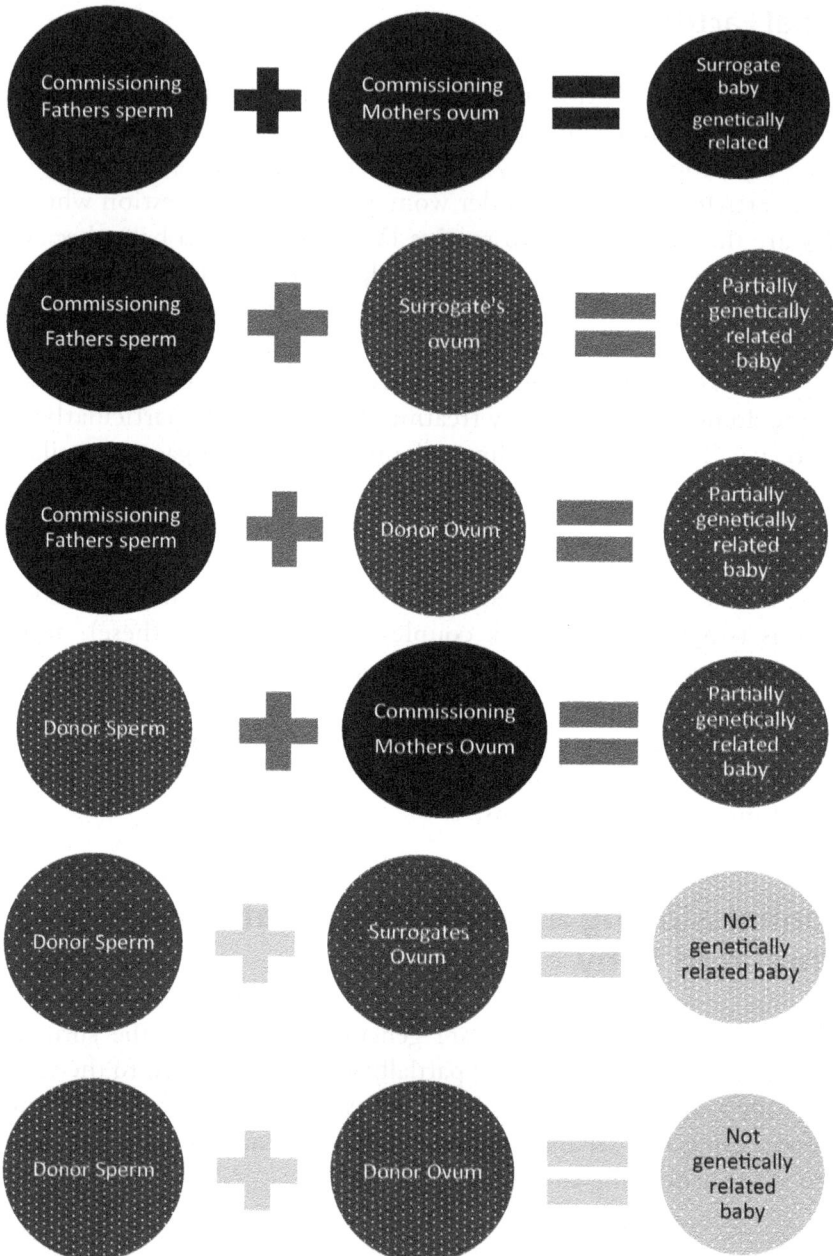

Fig. 5.2   Showing genetic link possibilities in the surrogate baby

Commissioning a Pregnancy and Delivery

Commissioning parents do not gestate or give birth to the baby they commission as this is the task of the surrogate mother. This fragmentation of reproductive function has consequences for all parties concerned. From the perspective of the commissioning parents, commissioning a baby through surrogacy guarantees a newborn baby into their family but it does separate sex from reproduction and bringing about the baby. It also separates motherhood from pregnancy which may have implications for later attachment with the baby, since no opportunity to form these bonds or attachments during pregnancy exists between them. Recent research suggests maternal-foetal sensitivity is associated with more maternal-baby sensitivity (Maas et al. 2016). There is also a general lack of breastfeeding opportunity for commissioning parents, be they men or women, although there are mothers who seek to breastfeed through therapeutic means (they usually receive hormones to mimic pregnancy and are encouraged to use a breast pump to stimulate and artificially induce lactation). Finally, as described in van den Akker (2000), commissioning a surrogate baby separates the unity of a couple into a threesome, the third party (the surrogate, her pregnancy and in many cases also donated gametes) being invited into this intimate process of reproducing. In the majority of cases this is much more than a third party since gestational surrogacy involves a whole clinical team (including, e.g. embryologists, clinicians, nurses and counsellors) necessary to bring the pregnancy about. These multiple involvements fragmenting the traditional ideal kinship ideology of what the family means have implications for the new family, their social network and society more generally.

The Family

From the perspective of the family, the commissioning individual man, woman or couple place the surrogate baby into the extended family of grandparents, aunts, uncles, nieces and nephews. The surrogate baby does not become 'theirs' until the baby is relinquished by the surrogate mother and formally taken into the family through the process in the United Kingdom of a PO (see Chap. 10). The new commissioning family is not certain of their legal status as responsible parents until that process is complete. For the entire family network, the baby's arrival (including the granting of the PO) therefore finally marks it as part of that unit, not before as is the case when a couple conceives and gestates their own baby. This 'gestational waiting' period can cause anxiety for the prospective family because there is no certainty that the baby will be theirs, even when their own gametes have been used (van den Akker 2000). The fact that the gestation period was externally conceived and that it was commissioned has led to discomfort as it calls for a need to redefine 'family'. Snowdon (1994) also referred to a 'presumed fragility of a family relationship not supported by genetic relatedness'. Non-gestational and non-genetically related families are therefore unorthodox. The traditional family is threatened and new forms of family are seen as deviant (Bernardes 1993, 2000). Redefinitions of family are not just due to the third-party involvement but also because of the changing face of society with single men and women, older men and women, same-sex and heterosexual divorced and step relationships now common and all seeking ways to become parents and form new families. This is discussed further in the chapter on parenting (Chap. 6) and the effects on children (Chap. 7).

Post-modern family theorists (see also Chap. 2) argue that with the current fluidity of the population and its partnership structures described above, there is little traditional about modern families. For example, with one in three marriages breaking up, and many of those embarking upon new partnership creating step-siblings and mixed genetic offspring within the newly constituted families, and single parenting and gay and lesbian marriages all increasing and seeking children within their households, what was unorthodox is rapidly becoming a new tradition (Bernardes 1993). The traditional structural functionalist family (Parsons 1959) was previously the normative, male/female married couple starting a biological family structure which was institutionalised and culturally legitimised. Interestingly new or post-modern family forms called for institutionalisation and cultural legitimisation as shown, for example, by the ample uptake of same-sex marriage (Marriage (same sex couples) Act (2013)) reaching 1409 between 29 March and 30 June 2014 (ONS 2014). This homogenised version of the non-traditional family does not serve the families well, because many are trying to turn a non-traditional family into a traditional one when they are not. In some cases, anonymous arrangements continue to be used in gamete and embryo donation and even in surrogate motherhood. This leaves the children conceived via these routes with unclear and uncertain genetic information and robs them of the potential development of their sense of self, failing connectedness to the family community and no bridge to past generations (van den Akker 2001a). These fears for the children are doubly devastating for those commissioned and subsequently rejected by the parents who commissioned their existence and subsequently reneged on taking the baby (Evans 2015; ABC News 2016). From a research, theory and practice perspective, any traditional notions of the family should be rejected in third-party-assisted conception, as they fail to meet the needs for traditional family category analysis and comparison (Bernardes 1993), in favour of a non-traditional post-modern framework, where this unusual reproduction is given its own legitimate existence.

Cognitive Dissonance

The theoretical shift from traditional to post-modern family has not yet fully been adapted by the populations involved in creating this shift in practice nor has it been fully adapted by the communities in which these new families live (van den Akker 2000, 2001a, 2016). This lack of theory to practice adaptation to change has led to a discord between what people do and what they believe in, referred to as cognitive dissonance. For example, when couples seek treatment for medical or social infertility or involuntary childlessness, they seek a preferred option mimicking as close to the traditional family as possible. If a couple can gestate the pregnancy, they prefer that to sourcing out the gestation; if they can use their own gametes, they prefer that to using donated gametes; and if they can commission a surrogate baby, they prefer that to adoption (with a presumed total absence of genetic and gestational input) (van den Akker 2007a). van den Akker (2000, 2001a, b) reported on the likely cognitive dissonance experienced by infertile mothers. In van den Akker (2000), 24% of all commissioning mothers studied chose surrogacy to try to have a genetic link with the baby. However, of the commissioning mothers using genetic surrogacy 31% believed a genetic link was important, even though they could not use their own ova in the surrogate arrangement. They were also asked if a genetic link was important to their partner and 69% said they believed their partners thought a genetic link with the mother was important, yet they did not have this maternal genetic link in the resultant family. This shows that even couples embarking upon a surrogate arrangement where there would be no maternal genetic link believed this to be important. In the 2001b study, van den Akker reported infertile heterosexual couples were least likely to disclose third-party-assisted conception to their child. Also, those who believed a genetic link was important were less likely to disclose genetic difference with the child.

Little research has yet studied the effects of cognitive dissonance in third-party reproduction, and only time will tell if this is problematic within the family in the longer term. It is however known that individuals generally strive for cognitively consonant or harmonious states, as disequilibrium between actions and beliefs are known to lead to discomfort (Festinger 1957). In Blyth's (1995) study of British commissioning parents, no denial of the fact that they used surrogacy took place suggesting full cognitive consonance, as was found in van den Akker's (2000) research. However, genetic difference was not always disclosed. In Ragone's (1994) research, more non-disclosure of surrogacy was reported. The differences between British and American studies may be because the practice is different in the United States, with surrogacy there being more commercially orientated than British surrogacy. The transition between infertility resolution and readiness to (adopt or) opt for third-party-assisted conception including surrogate motherhood is not always achieved by all individuals. This process of negotiation of alternative avenues to traditional parenthood has been called 'mazing' (Sandelowski et al. 1989). Mazing involves drawing upon cognitive processes, comparisons, justifications, hope and so on, ultimately, if successful, leading to cognitive reframing or consonance with the outcome. This process of cognitive restructuring is made easier if alternatives can be compared favourably (van den Akker 2001a). As in adoption, when this process is successfully completed the new parent(s) of the child need to integrate this adaptation into the extended family and community and share the genetic/gestational difference with the child. However, prospective parents using third-party conception have more to face, such as their social or medical infertility, and loss of biogenetic connection and history with the child. If there is no appropriate cognitive adaptation and resolution, optimum communication is unlikely to happen (Cook et al. 1995) and pathogenic relationships can result, as is found in adoption (Brinich 1990; Toussieng 1962). Moreover, the future advances and uses of genetic testing are likely to 'drive anonymous gamete donation (and therefore also anonymous surrogate motherhood or surrogacy using anonymous gametes or embryos) out of business' (see challenging article by Harper et al. 2016).

The public acknowledgement of the family being 'different' remains important once the baby arrives in the new family. In 2002, we reported on a discrepancy in birth registrations, with commissioning parents not always adhering to the process of POs. In our study (Crawshaw et al. 2012) there was a discrepancy between surrogate births recorded by surrogacy agencies and General Register Offices (GRO) in the United Kingdom between the years 1995 and 2011. Only half of the commissioning parents actually proceeded with a PO application, suggesting some commissioning parents do not have legal parentage rights or responsibilities for their surrogate-born babies or that they incorrectly registered these births as their own. From 2011, the picture reverses with slightly more GRO registered surrogate births than from UK surrogate agencies. This discrepancy is likely to be explained by the increase in cross-border surrogate births. According to the GRO, 26% of surrogate births took place abroad in 2011. Inaccuracies in birth registrations are therefore apparent and have consequences for family functioning and the child's rights to correct information about his or her origins. The births which are registered for example do not provide any information about the residence of the surrogate mother that was used in its conception, an omission that urgently needs to be corrected.

Disclosure

Considering that family building using gamete donation has increased substantially over the last decade, it is surprising that the attitudes towards disclosure remain difficult. In the United Kingdom alone in 2013, there were 48,477 fresh IVF cycles and 4611 cycles using donor insemination.

Of the 2013 fresh IVF cycles, 5% used donor sperm, 4% used donor eggs and <1% used both donor eggs and sperm, or donor embryos—totalling just under one in 10 treated cycles using donor gametes, showing widespread use of donated gametes/embryos (HFEA 2014). Interestingly the HFEA also has data available for patients accessing fertility treatment, with of the 42,721 (fresh IVF and DI cycles) in 2013, 1015 (2%) had no registered partner and 1342 (3%) registered with a female partner (HFEA 2014), showing the family structures using gamete donation are often non-traditional too. Despite this, disclosure remains an issue.

Issues of disclosure are closely related to the amount of comfort or discomfort (or cognitive dissonance) individuals feel in relation to third-party family building. Research on gamete donation has shown that secrecy is paramount and continues to dominate the practice internationally. Secrecy is traditionally the result of advice given by clinicians to pretend the child is genetically theirs because in gamete donation, a pregnancy in heterosexual couples can hide the third-party involvement of eggs, sperm or a donated embryo. In surrogacy this is not possible as there is no pregnancy and it is much more difficult to hide that fact from family, friends and their community. Nevertheless, when asked if they would disclose a number of third-party options to build a family, van den Akker (2000) found that all but one commissioning mother would disclose a surrogacy arrangement to their surrogate child. However, the questions were also broken down by those women who could use their own oocytes and those who could not. Of those who could use their own oocytes, all but one would disclose surrogacy, whereas all women who needed to use genetic surrogacy said they would disclose using surrogacy to the child. However, the picture gets more complicated. As can be seen below, 16 women could use their oocytes in the gestational surrogate pregnancy and slightly fewer would disclose the use of IVF, sperm or egg donation to their child, but most would disclose adoption if the child had been adopted (see Fig. 5.3). On the other hand, of the 13 women who were not able to use their own genes in the (genetic) surrogate pregnancy, slightly fewer were likely to tell their child if IVF, sperm or egg donation was used, and only about half would disclose adoption (see Fig. 5.4).

This demonstrates substantial cognitive dissonance even in commissioning parents who are themselves all using one of the most complicated

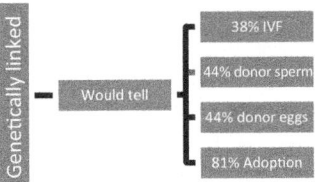

Fig. 5.3   Showing disclosure of IVF, sperm or egg donation or adoption

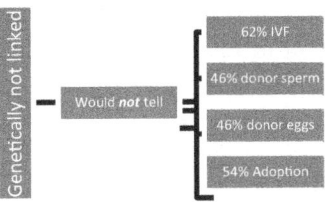

Fig. 5.4   Showing non-disclosure of IVF, sperm or egg donation or adoption

routes to becoming a family. Their dissonance may have potential implications for the children as they are conceived via the less preferred option (see van den Akker 2007a) or the option they are most secretive about. Either way, these children will not be told their true genetic origins. Interestingly, those with a genetic link in a surrogate baby were more likely to disclose IVF use in a child, but less likely to disclose adoption, showing how closely guarded they are about genetic links. Conversely, those who did not have a genetic link with the child were more likely to disclose adoption but not IVF use, showing they did not closely guard a lack of genetic link, although why they would not reveal IVF use is not clear. These two groups of women were similar in their willingness to disclose gamete donor use.

In Ragone's (1994) American research on commissioning mothers, individuals were found to emphasise and disclose selective aspects which mimic traditional family ideologies, and they tended to ignore those which differed. Blyth (1995) on the other hand reported that individuals did not deny difference in their attempts to normalise their use of surrogacy in his British research. In gamete donation research it is known that parents frequently do not disclose to their children they were conceived via assisted conception. Tallandini et al. (2016) reported a probability of less than 50% disclosure to children under the age of 10 years. They fear that disclosing assisted conception may undermine their parent child relationship and affect the child's development negatively (MacCallum and Golombok 2007; Readings et al. 2011; Sälevaara et al. 2013). The reality is that these fears are unfounded (Freeman and Golombok 2011; Daniels et al. 2011). The consequences of non-disclosure rob the children of their right to accurate genetic health and illness information and are put at risk of potential consanguineous relationships. It also puts them at risk of finding out later in life accidentally or via the increasing use of DNA testing—when they may be ill-prepared to process the implications.

Concerns About the Surrogate Arrangement

Clearly it takes time for commissioning couples to find a suitable surrogate, with some needing to negotiate an agreement with several surrogates before a commitment to pursue the arrangement is made (van den Akker 2000). In that study, nearly half the commissioning mothers expected a good relationship with the surrogate and about a quarter believed it would be difficult for a surrogate to relinquish the baby. This is expected to add to the tension of an arrangement with an already uncertain outcome. Very few were concerned about the expenses they would incur in commissioning the arrangement and social stigma or had concerns that the commissioning gametes were not used (a strategy useful in de-emphasising the virtues of genetic over social parenting). Instead, if they had concerns, these were somewhat more focused on IVF failure, non-relinquishment of the baby, emotional difficulties for the surrogate mother, legal obstacles and common baby worries.

Becoming a Parent of a Surrogate Baby

There is no doubt that interventions and arrangements designed to help people becoming parents are intended to give them a much-longed-for infant. This infant commissioned into the intended parent(s) family enters a non-traditional environment. A detailed description of separation, loss and attachment is provided in the chapter on parenting (Chap. 6). In this chapter, the focus is on different aspects of the parenting experience. Questions have been asked about the new commissioning parents' coping strategies with the infant within the wider social environment, similar to questions previously asked in adoption. Questions historically asked include:

- Does she/he perceive the absence of a maternal-foetal gestational bond?
- Does she/he perceive the absence of a genetic bond?
- Does his/her identity change to biologically removed parent?
- Does the parent's own experience of loss and separation affect this new relationship?
- Are the parents' feelings of 'belonging' and acceptance/rejection different?
- Is there guilt about the mode of conception and taking the child from its biological parent(s)?
- Does the new parent perceive stigma, at what they have done to become new parents?
- Do the child's genetic characteristics (if different from their own) affect preferences and acceptance of these differences?
- Do they emphasise similarity and ignore difference?
- Hide or fabricate birth or genetic origins?
- Fear the reacting of the child if it finds out its genetic or gestational origins?

Commissioning parents do not benefit from the maternal-foetal bond that develops in preparation for the birth and post-natal attachment formations. This relationship is extremely important for the development of the child. Research (Cannella 2004) and theory (Raphael-Leff 1991; Bielawska-Bartorowicz 2006) have demonstrated that the attachment relationship generally starts during pregnancy. The important question of the need for a biological (gestational) connection between a parent and child and subsequent adequacy of parenting has been posed (e.g. Schuker 1988; Maas et al. 2016). Schuker suggests the importance of that biological connection depends more on the cognitive meaning attached to that biological bond than the bond itself. However, within the wider social context, parents and even grandparents or other family members may reject a non-biological child within the family (Blum 1983), and conversely, they may mourn the loss of a biological child if they are members of a genetic surrogate mother. However, according to Maas et al. (2016) the biological bond or maternal-foetal sensitivity experienced during pregnancy is associated with more maternal-baby sensitivity, suggesting it may be more important than previously thought.

Despite this, commissioning mothers (parents) of surrogate babies may not need the opportunities of direct, in vivo physiological bonding with the foetus because she/he may benefit from the interaction with the surrogate mother and bond via her with the baby. One study of commissioning mothers found they were much more likely to express concerns and worries about the babies than the surrogate mothers did (van den Akker 2007a). Similarly, fathers are also reported to be more attached to the foetus than the pregnant mothers (White et al. 1999), suggesting the physiological bond may not be the only factor facilitating prenatal attachment to the foetus. In a study comparing Swedish and Polish mothers, the strength of attachment as measured by total scores on the Prenatal Attachment Inventory were lower in Swedish mothers, but their conceptualisation of their relationship with the unborn child (reflected in fantasising about the baby, attributing traits, interacting with the unborn child and expressing affection for and allowing others to learn about the foetus) were similar (Bielawska-Batorowicz and Siddiqui 2008). Defined in this way, the majority of these attributes can apply to individuals lacking the gestation (men and commissioning mothers) too. According to Bielawska-Batorowicz and Siddiqui (2008), the factors they obtained from their data suggest a multidimensional structure to attachment, similar to structures proposed in other research across cultures. This could also explain the ability of some women to attach (as in commissioning parents) or detach (as in surrogate mothers) from the foetus (van den Akker 2007a, b), regardless of the physiological opportunities to bond, as this may be useful but not necessary for the development of attachments with the foetus.

It is well known that traditionally, adoptive parents deny birth differences with the child and fail to discuss its origins, thereby behaving 'as if' the child was genetically and gestationally their own. Similarities with surrogacy have been reported. Ragone (1996, p. 352), for example, describes the surrogate mother process from the perspective of the intended couples as 'focusing less on the aspects of surrogacy that depart from tradition, than they do on those that are consistent with American kinship ideology, most notably in their emphasis on the importance of family and nurturance'. Levine (2008) argues that kinship models created by some nontraditional families use conventional as well as radical ideas to reference biogenetic connections. This is evidenced in research where people coped with cognitive dissonance of the biogenetic distance with the child by cognitively restructuring new interpretations of third-party-assisted conception families (van den Akker 2007a). However, some parents are changing or reconstructing what is important to not important; what is legal to illegal and hiding/denying important facts. In Ragone's (1996, p. 359) research, intended mothers were quoted as saying 'Ann is my baby; she was conceived in my heart before Lisa's [the surrogate's] body'; and 'She [the surrogate] represented that part of me that couldn't have a child', representing, according to Ragone, mythical conceptions and symbiotic relationships respectively. The surrogate mothers in this work actively assist in the restructuring and removing dissonant tensions by actively de-emphasising the importance of gestation but also of the genetic bond they have with the baby, for example, 'She [the intended mother] was emotionally pregnant, and I was just physically pregnant' (Ragone 1996, p. 360).

Moral and Ethical Concerns

All parties involved in surrogate motherhood arrangements deliberately seek to bring about a child and this very fact—involving numerous parties—allows for a moral obligation to protect the unborn child from potential harm. The potential threat to the child in surrogacy is compounded by the fact that surrogate babies are commissioned and created via the surrogate mother specifically for the commissioning person(s). Any threat to the welfare of the child may not be known to the surrogate mother or to the commissioning parents. Although rare, there is some documented evidence of commissioning parents abusing surrogate-born children—using surrogacy as the route to the children. This chilling reality was published in a recent News Report (2016), which revealed how an Australian (Victorian) man admitted sexually abusing his baby surrogate twin daughters. He intentionally brought them to Australia to use for sex. Assisting a sick and criminal individual to achieve such hideous goals goes against all acceptable practices, suggesting some form of screening as purported decades ago is still relevant today. Mental health professionals' screening of prospective commissioning parents for a history of child abuse, neglect, cruelty or paedophilia is therefore desirable. However, cross-border arrangements make this virtually impossible as indicated by the complicated and well-described 'baby Gammy' case described earlier, (see Blackburn-Starza 2014). The intended biological father was a convicted sex offender (he was sentenced to three years in prison in 1997 for molesting two girls aged 7 and 10) but this was not known in Thailand where the surrogate arrangement took place (Hawley 2014). Screening could therefore act as a guarantor to the surrogate mother that the welfare of the baby she relinquished is paramount.

Kinship

Research on mothering and/or fathering a surrogate child has focused on intended commissioning mothers, although some research asked the surrogate and intended mothers how their partners felt (van den Akker 2003, 2007b). There are some notable exceptions of research on gay surrogate fathering (Bergman et al. 2010; Patterson and Riskind 2010; Dempsey 2013; Golombok et al. 2017). Ragone's (1996) ethnographic research questioned (heterosexual) fathers about their perceptions of fathering a surrogate baby. She quotes fathers as saying 'here she [the surrogate] is carrying my baby. Isn't she supposed to be my wife?'; 'I felt weird about another woman carrying my baby' and 'seeing Jane [the surrogate] in him [his son], it's literally a part of herself she gave. That's fairly profound.' The relationship between the fathers and surrogates is non-sexual but they produce a child together leading to awkwardness and concerns about how to behave. The father is also acutely aware that this child is not genetically or gestationally related to his wife which could result in conflict. In order to maintain some psychological balance, they often de-emphasise the role of the father, and that of the surrogate mother, reconstructing the importance of the nurturing and caregiving role. When possible, for example in national surrogacy arrangements, they also tend to emphasise the important 'bond' that has formed between the surrogate mother and the commissioning intended mother, as that presents a feeling and perception of unity rather than exclusivity to the arrangement, thereby validating the importance of the intended mother. Interestingly, few intended couples take up the offer of paternity testing, potentially fulfilling yet another function in reinterpreting and legitimising the new family. The commissioning mother may feel equalised by the possible lack of her husband's genetic link with the baby, and allowing paternal uncertainty can redress the symbolic imbalance created between the couple via the surrogate arrangement (Ragone 1996). According to Ragone, the three key persons (father, surrogate mother, commissioning mother) building this new family 'work together in concert to create a new idea of order and appropriate relations and boundaries by directing their attention to the sanctity of motherhood as it is illustrated in the surrogate and adoptive mother bond' (p. 361). By not confirming paternity, they are all downplaying the importance of the lack of genetic link in the mother. Hence the new family is socially, rather than genetically, constructed. This research, as the publication dates confirm, largely drew on genetic surrogacy. The case for gestational surrogacy is different. There both parents can have an equal genetic link with the child, unless donor gametes or embryos are used, which poses yet another scenario. Infertile commissioning parents, like their IVF counterparts across the world, have only limited knowledge of the procedures involved (see, e.g. Sohrabvand and Jafarabadi's 2005 study of Iranian couples' knowledge of assisted conception), and this only develops as they progress through the stages.

Summary

In summary, research focusing on commissioning parents is relatively rare and relatively incomplete. Surrogate motherhood is used by individuals and couples for medical and social reasons, either within their home country (if legally allowed) or abroad. The opportunities are limited to individuals or couples who can afford to pay for surrogate motherhood services and medical or financial factors may dictate the use of own versus the surrogate's or donated gametes. Inequality is therefore prevalent and this has led to concerns of the commodification of babies. Research and theory into the effects of a surrogate build family upon the extended family is virtually non-existent and sociocultural attitudes may determine the amount of cognitive dissonance or harmony the commissioning parents experience once they bring the surrogate-born infant into their home. Their quest to build a family through the surrogate motherhood route is therefore a complicated one, marked by vulnerabilities, insecurities, financial investments and a social context that leads some of them to deny that their unusual family is in any way different from a traditional family.

References

ABC News. (2016). http://www.abc.net.au/news/2016-04-14/baby-gammy-twin-must-remain-with-family-wa-court-rules/7326196. Accessed 7 Oct 2016.

Andersen, A., Gianaroli, L., Felberbaum, R., de Mouzon, J., & Nygren, K. (2005). Assisted reproductive technology in Europe, 2001. Results generated from European registers by ESHRE. *Human Reproduction, 20*, 1158–1176.

Bergman, K., Rubio, R., Green, R., & Padron, E. (2010). Gay men who become fathers via surrogacy: The transition to parenthood. *Journal of LGBT Family Studies, 6*(2), 111–141.

Bergner, A., Beyer, R., Burghard, F., Klapp, F., & Rauchfuss, M. (2008). Pregnancy after early pregnancy loss: A prospective study of anxiety and depressive symptomatology and coping. *Journal of Psychosomatic Obstetrics and Gynecology, 29*(2), 105–113.

Bernardes, J. (1993). Responsibilities in studying post-modern families. *Journal of Family Issues, 14*, 35–49.

Bernardes, J. (2000). *Family studies: An introduction* (pp. 27–46). London: Routledge.

Beutel, M., Wilner, H., Deckhart, R., Von Rad, M., & Weiner, H. (1996). Similarities and differences in couples' grief reactions following miscarriage: Results from a longitudinal study. *Journal of Psychosomatic Research, 40*, 245–253.

Bielawska-Bartorowicz, E. (2006). *Psychologiczne aspekty prokreacji* [Psychological aspects of human reproduction]. Katowice: 'Slask' sp. Z o.o. Wydawnictwo Naukowe.

Bielawska-Batorowicz, E., & Siddiqui, A. (2008). A study of prenatal attachment with Swedish and Polish expectant mothers. *Journal of Reproductive and Infant Psychology, 36*(4), 373–384.

Blackburn-Starza, A. (2014). Surrogacy: Confusion amid reports of boy 'abandoned' by intended parents. *BioNews*, p. 765. http://www.bionews.org.uk/page_442973.asp. Accessed 29 Apr 2016.

Blum, H. P. (1983). Adoptive parents: Generative conflict. *Psychoanalytic Study of the Child, 38*, 141–164.

Blyth, E. (1995). Not a primrose path: Commissioning parents experiences of surrogacy arrangements in Britain. *Journal of Reproductive and Infant Psychology, 13*, 185–196.

Boivin, J. (2003). A review of psychosocial interventions in infertility. *Social Science & Medicine, 57*, 2325–2341.

Brinich, P. M. (1990). Adoption from the inside out: A psychoanalytic perspective. In D. Brodzinsky & M. Schechter (Eds.), *The psychology of adoption*. New York: Oxford University Press.

Brodzinsky, D. (1997). Infertility and adoption adjustment: Considerations and clinical issues. In S. R. Leiblum (Ed.), *Infertility, psychological issues and counseling strategies*. New York: Wiley.

Caglar, G., Asimakopoulos, B., Nikolettos, N., Diedrich, K., & Al-Hasani, S. (2005). Preimplantation genetic diagnosis for aneuploidy screening in repeated implantation failure. *Reproductive Biomedicine Online, 10*, 381–388.

Cannella, B. L. (2004). Maternal-foetal attachment. *Journal of Advanced Nursing, 50*, 60–68.

Cook, R., Golombok, S., Bish, A., & Murray, C. (1995). Disclosure of donor insemination: Parental attitudes. *American Journal of Orthopsychiatry, 65*, 549–559.

Corbet-Owen, C., & Kruger, L. M. (2001). The health system and emotional care: Validating the many meanings of spontaneous pregnancy loss. *Family Systems Health, 19*(4), 411–427.

Crawford, N., Hoff, H., & Mersereau, E. (2017). Infertile women who screen positive for depression are less likely to initiate fertility treatments. *Human Reproduction, 32*(3), 582–587.

Crawshaw, M., Blyth, E., & van den Akker, O. (2012). The changing profile of surrogacy in the UK – Implications for national and international policy and practice. *Journal of Social Welfare and Family Law, 34*, 1–11.

Daniels, K. R., Grace, V., & Gillett, W. (2011). Factors associated with parents' decisions to tell their adult offspring about the offspring's donor conception. *Human Reproduction, 26*, 2783–2790.

Daugirdaite, V., van den Akker, O. & Purewal, S. (2015). Posttraumatic stress and posttraumatic stress disorder after termination of pregnancy and reproductive loss: A systematic review. *Journal of Pregnancy, 2015*, 646345, 14 pages.

de Liz, T., & Strauss, B. (2005). Differential efficacy of group and individual/couple psychotherapy with infertile patients. *Human Reproduction, 20*, 1324–1332.

Dempsey, D. (2013). Surrogate gay male couples and the significance of biogenetic paternity. *New Genetics and Society, 32*(1), 37–53.

Evans, A. (2015, November 26). *Sorry Sherri Shepherd, but you can't renege your baby's surrogacy*. LawStreetmedia.com. http://lawstreetmedia.com/blogs/entertainment-blog/sorry-sherri-shepherd-cant-renege-babys-surrogacy/. Accessed 5 Oct 2016.

Festinger, L. (1957). *A theory of cognitive dissonance*. Stanford: Stanford University Press.

Freeman, T., & Golombok, S. (2011). Donor insemination: A follow up study of disclosure decisions, child adjustment and family relationships at adolescence. *Reproductive Biomedicine Online, 25*, 193–203.

Golombok, S., Blake, L., Slutsky, J., et al. (2017). Parenting and the adjustment of children born to gay fathers through surrogacy. *Child Development*. doi:10.1111/cdev.12728.

Harper, J. C., Kennett, D., & Reisel, D. (2016). The end of donor anonymity; how genetic testing is likely to drive anonymous gamete donation out of business. *Human Reproduction, 31*(6), 1135–1140.

Hawley, S. (2014, September 17). Baby Gammy story takes startling turn as extreme options revealed. Australian Broadcasting Corporation. Archived from the original on 19 September 2014. http://www.abc.net.au/7.30/content/2014/s4089822.htm. Accessed 22 June 2017.

Haws, R. A., Mashasi, I., Mrisho, M., Schellenberg, J. A., Darmstadt, G. L., & Winch, P. J. (2010). "These are not good things for other people to know" how rural Tanzanian women's experiences of pregnancy loss and early neonatal death may impact survey data quality. *Social Science & Medicine, 71*, 1764–1772.

Heinicke, C. M. (1995). Determinants of the transition to parenting. In M. Bornstein (Ed.), *Handbook of parenting: Vol 3. Status and social conditions of parenting*. Mahwah: Erlbaum.

HFEA. (2014). http://www.hfea.gov.uk/docs/Egg_and_sperm_donation_in_ the_UK_2012-2013.pdf. Accessed 28 Apr 2016.

Hsu, M., Tseng, Y., & Kuo, L. (2002). Transforming loss: Taiwanese women's adaptation to stillbirth. *Journal of Advanced. Nursing, 40,* 387–395.

Jayakrishnan, K. (2012). *Insights into infertility management* (p. 266). New Delhi: Jaypee Brothers Medical Publishers.

Levine, N. (2008). Alternative kinship, marriage and reproduction. *Annual Review of Anthropology, 37,* 375–389.

Maas, et al. (2016). A longitudinal study on the maternal-fetal relationship and postnatal maternal sensitivity. *Journal of Reproductive and Infant Psychology, 34*(2), 110–121.

MacCallum, F., & Golombok, S. (2007). Embryo donation families: Mothers' decisions regarding disclosure of donor conception. *Human Reproduction, 22,* 2888–2895.

Margalioth, E., Ben-Chetrit, A., Gal, M., & Eldar-Geval, T. (2006). Mini review – Developments in reproductive medicine. Investigation and treatment of repeated implantation failure following IVF-ET. *Human Reproduction, 21*(12), 3036–3043.

Marriage (same sex couples) Act. (2013). http://webarchive.nationalarchives. gov.uk/20160105160709/http://www.legislation.gov.uk/ukpga/2013/30/ contents. Accessed 28 Apr 2016.

News Report. (2016, April 26). Man pleads guilty to sexually abusing surrogate twin baby daughters, court hears. *The Guardian.* https://www.theguardian. com/australia-news/2016/apr/29/man-pleads-guilty-to-sexually-abusing-surrogate-twin-baby-daughters-court-hears

Office for National Statistics. (2014). http://webarchive.nationalarchives.gov. uk/20160105160709/http://www.ons.gov.uk/ons/rel/vsob1/marriages-in-england-and-wales--provisional-/for-same-sex-couples-q1-and-q2-2014/sty-same-sex-marriages.html. Accessed 28 Apr 2016.

Parsons, T. (1959). The social structure of the family. In R. Anshen (Ed.), *The family: Its function and destiny.* New York: Harper & Row.

Patterson, C., & Riskind, R. (2010). To be a parent: Issues in family formation among gay and lesbian adults. *Journal of LGBT Family Studies, 6*(3), 326–340.

Payne, N., & van den Akker, O. (2016, May). Infertility network UK survey on the impact of fertility problems. Report commissioned by and produced for INUK.

Perla, L. (2001). Is in-vitro fertilization for older women ethical? A personal perspective. *Nursing Ethics, 8*(2), 152–158.

Ragone, H. (1994). *Surrogate motherhood: Conception in the heart* (pp. 114–120). San Francisco: Westview Press.

Ragone, H. (1996). Chasing the blood tie: Surrogate mothers, adoptive mothers and fathers. *American Ethnologist, 23*(2), 352–365.

Raphael-Leff, J. (1991). *Psychological processes of childbearing*. London: Chapman and Hall.

Rawlinson, K. (2014). http://www.theguardian.com/lifestyle/2014/aug/23/interpol-japanese-baby-factory-man-fathered-16-children. Accessed 15 Apr 2016.

Raziel, A., Friedler, S., Schachter, M., Strassburger, D., & Ron-El, R. (2000). Successful pregnancy after 24 consecutive fetal losses: Lessons learned from surrogacy. *Fertility and Sterility, 74*(1), 104–106.

Readings, J., Blake, L., Casey, P., Jadva, V., & Golombok, S. (2011). Disclosure and everything in between: Decisions of parents of children conceived by donor insemination. *Reproductive Biomedicine Online, 22*, 485–495.

Rosenfeld, J. (1991). Bereavement and grieving after spontaneous abortion. *American Family Physician, 43*, 1679–1684.

Sälevaara, M., Suikkari, A. M., & Söderström-Anttila, V. (2013). Attitudes and disclosure decisions of Finnish parents with children conceived using donor sperm. *Human Reproduction, 28*, 2746–2754.

Sandelowski, M., Harris, B., & Holditch, D. (1989). Mazing: Infertile couples and the quest for a child. *Journal of Nursing School, 21*, 220–226.

Schuker, E. (1988). Psychological effects of the new reproductive technologies. *Women and Health, 13*(1–2), 141–147.

Smolin. (2016). Surrogacy as the sale of children, Pepperdine. *Law Review, 43*, 265–311.

Snowdon, C. (1994). What makes a mother? Interviews with women involved in egg donation and surrogacy. *Birth, 21*, 77–84.

Sohrabvand, F., & Jafarabadi, M. (2005). Knowledge and attitudes of infertile couples about assisted reproductive technology. *Iranian Journal of Reproductive Medicine, 3*(2), 90–94.

Tallandini, M. A., Zanchettin, L., Gronchi, G., & Morsan, V. (2016). Parental disclosure of assisted reproductive technology (ART) conception to their children: A systematic and meta-analytic review. *Human Reproduction, 31*(6), 1275–1287.

Taranissi, M., El-Toukhy, T., Gorgy, A., & Verlinsky, Y. (2005). Influence of maternal age on the outcome of PGD for aneuploidy screening in patients with recurrent implantation failure. *Reproductive Biomedicine Online, 10*, 628–632.

Toussieng, P. W. (1962). Thoughts regarding the etiology of psychological difficulties in adopted children. *Child Welfare, 41*, 59–65.

van den Akker, O. B. A. (2000). The importance of a genetic link in mothers commissioning a surrogate baby in the UK. *Human Reproduction, 15*(8), 110–117.

van den Akker, O. B. A. (2001a). The acceptable face of parenthood: Psychosocial factors of infertility treatment. *Psychology Evolution and Gender, 3*(2), 137–153.

van den Akker, O. B. A. (2001b). Adoption in the age of reproductive technology. *Journal of Reproductive and Infant Psychology, 19*(2), 147–159.

van den Akker, O. B. A. (2003). Genetic and gestational surrogate mothers' experience of surrogacy. *Journal of Reproductive and Infant Psychology, 21*(2), 145–161.

van den Akker, O. B. A. (2007a). Psychosocial aspects of surrogate motherhood. *Human Reproduction Update, 13*(1), 53–62.

van den Akker, O. B. A. (2007b). Psychological trait and state characteristics, social support and attitudes to the surrogate pregnancy and baby. *Human Reproduction, 22*(8), 2287–2295.

van den Akker, O. B. A. (2011). The psychological and social consequences of miscarriage. *Expert Review of Obstetrics and Gynecology, 6*(3), 295–304.

van den Akker, O. B. A. (2012). *Reproductive health psychology*. Wiley-Blackwell. ISBN-13: 978-0470683385.

van den Akker, O. B. A. (2013). For your eyes only: Bio-behavioural and psychosocial research objectives. *Human Fertility, 16*(1), 89–93.

van den Akker, O. (2016). Reproductive health matters. *The Psychologist, 29*(1), 2–5.

White, M. A., Wilson, M. E., Elander, G., & Persson, B. (1999). The Swedish family: Transition to parenthood. *Scandinavian Journal of Caring Sciences, 13*, 171–176.

6

Separation and Parenting a Surrogate Baby

This chapter discusses traditional conceptions where parenting rights and responsibilities usually belong to the parents conceiving the baby. Traditionally, separation and loss of a baby is traumatic and parenting is assumed to be natural. These contexts are also rooted in national policies and practices which in most cases do not facilitate the social and lifestyle changes and technological and economic developments in family building through third-party reproduction. Biological, genetic, psychological, social and legal parenthood is fragmented and may affect the welfare of the child. Bonding and forming attachments are positive for all parent-child relationships, and gaps in the opportunity to form these attachment bonds, or an absence of bonding opportunities or separation can have long-lasting effects on these critical relationships and, later, child behaviours.

In surrogate motherhood, the question who the biological, genetic, psychological, social and legal parent of the surrogate commissioned baby is, is not straightforward. Contractually parenthood goes to the intended parent(s) but legal systems differ across countries so legal parenthood is not universally the same. For example, in the United Kingdom legal parenthood is automatically the birth mother, whereas in India it is the genetic parent. According to biological and genetic perspectives

© The Author(s) 2017
Olga B.A. van den Akker, *Surrogate Motherhood Families*,
DOI 10.1007/978-3-319-60453-4_6

the biological, gestating, birthing and genetic mother is the parent. In psychological terms, the parent who nurtures and bonds with the foetus in utero and socially the raising parent(s) could be considered the parent(s) of a surrogate baby. Parenthood may therefore differ by psychosocial factors, culture and country. Table 6.1 indicates the potential rights and responsibilities of the possible parents after a surrogate motherhood arrangement. These likely, possible and unlikely parental rights and responsibilities could apply to all countries participating in third-party-assisted conception and surrogate motherhood. Depending upon the local sociocultural and legal emphasis on genetics, a gestational and

Table 6.1 Showing universally likely, possible and unlikely parental rights and responsibilities following surrogate motherhood arrangements

	Intended parent(s)		Donors in surrogacy	Surrogate mother	
	Using own gametes	Using surrogate or donor gametes	Gamete/ embryo	Gestational using commissioning or donated gametes	Genetic using own gametes
Pre-birth order	Yes	Yes	Possibly	No	No
Post-birth registration	No	No	No	Yes	Yes
Arrangement contract	Yes	Yes	No	No	No
Biological parental rights	No	No	No	Yes	Yes
Genetic parental rights	Yes	No	Yes	No	Yes
Psychological attachment to foetus	Possibly	Possibly	Possibly	Yes	Yes
Social parental responsibility to surrogate baby	Yes	Yes	Possibly	Possibly	Possibly
Moral obligation to offspring	Possibly	Possibly	Possibly	Possibly	Possibly

genetic surrogate mother in the United Kingdom, for example, may bond with the foetus in utero differently affecting her ability to separate herself from the developing foetus as her child (van Zyl and van Niekerk 2000). In countries such as Japan where little information on genetics is available to the population (Minai et al. 2007), this is not considered important in determining parenthood. Japanese surrogate mothers may therefore be expected not to find separation from a genetic or a gestational surrogate baby different and in all countries a surrogate mother may feel equally responsible for the social welfare of the child.

Commissioning parents have rights to reproduce and plan a baby just like any other parents, but the planning is uncertain because a surrogate mother may change her mind. There are calls in the United Kingdom for a change in legislation to allow intended commissioning parents to draw up a preconception agreement where the surrogate mother relinquishes her rights to parent that child even before it is conceived similar to gamete donors' relinquishment of rights to the child resulting from their donation (Robertson 1990). This proposal, which is narrow in scope, is gaining in popularity as evidenced by a recent report (Horsey 2015). What they fail to recognise is that any moral obligation to parent the child will depend upon the individuals' views of what it is contracting or donating. Donors are known to differ in how they see their gametes; they may attribute 'child' or 'cell' status to their donation (Purewal and van den Akker 2007). Surrogacy involves more than the donation of gametes; it involves a long process of conceiving and nurturing a growing unborn baby followed by a delivery and a handing over of a baby which was part of a woman (the surrogate mother) for nine months. Depending upon how important a genetic link is to a commissioning parent, she/he too may be equally or differently committed to parent a genetically or non-genetically related surrogate baby. From a parenting perspective, intending to become a parent of a baby carried by a surrogate mother is therefore an insufficient reason to determine parenthood. Van Zyl and van Niekerk (2000) argue that assuming intentionality as a prerequisite for parenthood is also morally unacceptable as it objectifies babies. Furthermore, psychologically it could harm a surrogate mother if she is unable to proceed with the relinquishment but is forced to separate from the baby because of a pre-birth contract.

Clearly legal, moral, ethical, social and psychological factors are important considerations in surrogate parenthood, particularly if the welfare of
the child is paramount. It is not useful to separate the relevance of each
of these factors (ethical, moral, legal, social, psychological) as they are
likely to be interconnected. Separating them by, for example, opting for
legally drawn-up pre-birth contracts and placing control of the foetus
during pregnancy in the hands of the intended commissioning parent(s)
makes the position for a surrogate mother psychologically, ethically and
morally unacceptable if they wish her to have (or not have) an abortion.
The surrogate mother's responsibility for the welfare of the child or her
personal feelings of having an abortion may oppose that of the intended
commissioning parent in the same way as when a non-surrogate woman
who conceives may not want to abort the foetus even if her partner did
not agree to a pregnancy. Here, neither the intended commissioning
parent(s) nor the father is deemed to be responsible for the upbringing
of that child. The surrogate mother may also feel personally unable to
relinquish the baby regardless of the genetic input, law or contract. A
fuller understanding of what makes a parent from research on parenting
and from a consideration of social changes in families and their functions
is particularly relevant in relation to the surrogate mother who nurtures
a developing foetus.

Non-traditional Families

The social context of being a child parented by commissioning parents of
a surrogate mother could also have implications for the child. Single parenthood was historically tainted with images of deprivation and 'need',
whereas parenting a child in a same-sex relationship was tainted with
images of children being bullied and stigmatised for having same-sex parents. The social acceptance of differences in family structures is improving, but not universal. Research has shown that children who experience
adverse reactions to being raised in a non-traditional family may become
socially withdrawn and have problems maintaining friendships, lower
self-esteem and potentially adverse psychological wellbeing in later life
(Mooney-Somers and Golombok 2000). Where no adverse reactions are

experienced, children of single and lesbian mothered families (Golombok et al. 1983) and children of lesbian and heterosexual families (Golombok and Tasker 1997) have similar psychosocial outcomes. This suggests that children brought up via surrogate arrangements are also likely to fare well if no adverse reactions to their non-traditional family are experienced. However, that is assuming the offspring are aware of the circumstances of their birth. In a study of 33 genetic and gestational surrogate families (Jadva et al. 2012), 30/33 children aged 10 were told about their surrogate conception, and 3 (two gestational-known and one gestational anonymous) were still planning to tell their child at a later time. Where genetic surrogacy was used, all commissioning parents had informed their child about surrogacy by age 10 years. At age 10 years, 19 children who had been informed of the nature of their conception had a good understanding of this and 13 of the 14 children who were in contact with their surrogate reported that they liked her. Fourteen children aged 7–10 years had seen their surrogate mother in the past year and most were happy with their level of contact with her or would have liked to see her more. At age 10 years, most children (14, 67%) felt neutral or indifferent about being born through surrogacy (Jadva et al. 2012). In a recent study comparing gay and lesbian parent families, lower levels of children's internalising problems were reported by gay fathers. However across both types of family, if the gay or lesbian parents thought there were higher levels of stigmatisation, and where their children experienced more negative parenting, higher levels of parent-reported externalising problems were reported (Golombok et al. 2017).

Parenting

Attachment between a parent and an infant is essential for safe and secure adaptation to the environment and interactions within that environment and insecure attachment styles are a risk factor for later psychopathology (Mikulincer and Shaver 2012). Early interpersonal attachments therefore constitute fundamental human motivations and emotions with lasting effects into adulthood. The attachment between a mother and a foetus is also important (van den Akker 2012; Maas et al. 2016) and a potential lack of attachment in surrogates during pregnancy or intended commissioning parents immediately post-delivery may affect a surrogate child's adaptation to his or her environment. A father's attachment to his baby in traditional families is generally understudied, yet they too are known to play a crucial role in effective fathering (Habib and Lancaster 2005) and preventing infant cognitive delay (Bronte-Tinklew et al. 2008). Paternal attachment with surrogate-born infants in heterosexual families is understudied. This is potentially problematic because having a surrogate baby is already a stressful process riddled with uncertainty right up until the moment the surrogate mother relinquishes the baby. In birth mothers and fathers parental stress has been shown to be associated with avoidance of attachment (see Jones et al. 2015 for a review). Wynter et al. (2016) studied paternal characteristics and father-infant attachment and reported significant associations between oversensitivity, mood and poorer quality partner relationship and partner criticism of infant care.

This is important because previous research into IVF treatment has shown that for couples using assisted conception the transition to parenting is different from traditionally conceiving parents. Most have an infertility problem and undergo multiple treatments over a lengthy period of time. The children are conceived outside the sexual relationship and IVF pregnancies present higher risks of medical complications including miscarriages, multiple births, prematurity and low birthweight (van den Akker 2012, 2015). Research on parent-child relationships and the child's psychosocial development in families with children conceived via IVF is inconclusive (e.g. Gibson et al. 2000a; Golombok et al. 2001; Colpin 2002). Some report no significant differences or positive effects of IVF parenting (Weaver et al. 1993; Golombok et al. 1995, 1996; van Balen 1996; Hahn and DiPietro 2001), less parenting stress (Golombok et al. 1995, 1996) and more competence (van Balen 1996) compared to non-IVF parents. Negative results for IVF families, such as being more overprotective than parents with naturally conceived children (Weaver et al. 1993; Hahn and DiPietro 2001); a lower sense of self-efficacy in caregiving and less autonomy-promoting behaviour compared with non-IVF mothers (McMahon et al. 1997); and lower scores on measures of socioemotional adjustment in school and higher scores on self-report measures of anxiety, aggression and depression compared with control children (Levy-Shiff et al. 1998) have also been reported. Parenting following assisted conception and surrogacy may therefore pose additional challenges which may affect parenting behaviours in different ways.

Separation

There is no real precedent specific to surrogate-born babies and separation or detachment from one parent and attachment to another, although there is some understanding of the impact of perceived loss from a biological mother in children. According to Moss and Moss (1975), in mother/child loss, without considering the dynamics of separation from the biological mother, it is difficult to understand the dynamics of reattachment to another mother. In surrogate motherhood the surrogate baby is usually, but not always, immediately removed from the surrogate mother who had a physiological bond with the baby and possibly a psychological or emotional maternal foetal attachment too (Ragone 1994; van den Akker 2003). Genetic surrogates who are gestating and delivering a baby who is as much genetically linked to her as her own children are, are more likely to report this than gestational surrogates who do not have a genetic link with the baby. No research has yet specifically determined differences in post-natal attachments between commissioning parents and genetic or gestational surrogate-born babies. Baslington (2002) argues that part of the detachment process is facilitated by the fact that payment takes place, but there is no information of payment potentially facilitating attachment in a commissioning parent. Within maternal-foetal detachment theory the surrogate mother distances herself from the growing foetus and the commissioning parents begins the mental preparation for the baby. Cross-border surrogates may hold and care for the baby a bit longer than is generally the case in national surrogate arrangements particularly if the intended parents need to book time off for flights and the journey to the country where the surrogate arrangement took place as was shown, for example, in the Panorama programme 'the House of Surrogates' (2013). Nothing is known about the effects of separation on a newborn baby from its surrogate birth mother because the questions have not yet been asked, but this separation from one could impact upon the commissioning parents' ability to attach to the baby or vice versa. The next sections explore these issues in more depth.

Attachment

Becoming a parent means an individual shifts from seeking protection and security to providing it. Individuals learn this in infancy, and these early experiences of secure attachment provide advantages in adulthood. Within a social cognition perspective representing the interplay between emotional cognitive processes and behaviours, a mother who is more 'avoidant of attachment' attributes negative distress to specific aspects of the baby, not the situation. Similarly, anxious mothers do not always recognise fear in their babies and instead tend to attribute the baby's distress to other factors, therefore not responding in synchrony with what the baby is communicating (Leerkes and Siepak 2006). It is possible that surrogates are anxious during the surrogate pregnancy because they carry the responsibility for maintaining an optimum uterine environment for the unborn baby. The commissioning parents too may be anxiously awaiting the surrogates' delivery of the baby and be anxious about becoming a parent because they lack the opportunity to bond during the pregnancy. The less effective, anxious caring behaviours of the (surrogate and commissioning) parents set the scene for the infant and child's own later attachment style. For example, the critically important infancy and childhood experiences including those gained during adolescence and adulthood determine the shape the next generation's parenting behaviour takes (see, e.g. Bowlby 1988, p. 5; Belsky 2005; De Carli et al. 2015).

Bonding and forming attachments are therefore positive for all parent-child relationships and gaps in the opportunity to form these attachment bonds, or an absence of bonding opportunities can have long-lasting effects on these critical relationships and on later child behaviours. Even in non-surrogate arrangements there may be parent-child separations, for example when a parent is hospitalised for a period of time without the baby, and these separations too may affect parent-child relationships later on. Bowlby's seminal research (1969/1982) has shown that for such periods of separation to have an effect no more than a week is needed

(Leventhal and Brooks-Gunn 2000). Research on institutional children who did not have opportunities to interact safely with responsive adults has shown they have severe emotional and social developmental problems (Tizard and Hodges 1978; Zeanah et al. 2005), and this was also true in temporarily fostered children (Lawrence et al. 2006). Secure attachment is so important that even foster children who are removed from their parents who maltreated them do worse than children who remain with their (maltreating) parents (Lawrence et al. 2006). Mental health problems are strongly linked to early separation and insecure or disorganised attachments in the early years (Moss et al. 2005), and extended separations of more than one month before the age of five years have been linked to personality disordered behaviours in adolescence and early adulthood by Crawford et al. (2009).

Attachment theory is based upon the principles of an infant's or a child's appraisal of the mother (or mother figure's) availability, with a lack of availability, meaning separation (temporary) or 'loss' (permanent) (Bowlby 1973), as shown in Fig. 6.1.

Later researchers considered the importance of lines of communication and certainty that the mother will respond to the communication (Ainsworth 1990) or maternal sensitivity (De Wolff and van Jizendoorn 1997) resulting in securely attached or insecurely attached children. Here early separations between birth and two years are critical, because at that stage a baby/child will not understand telephone accessibility of a mother or understand when she will return. After these early years, a child is more able to understand that a mother too has activities she needs to participate in, which do not involve the child, and the relationship between the mother and child is described as a 'goal corrected partnership'. Here open lines of communication allow the child to see

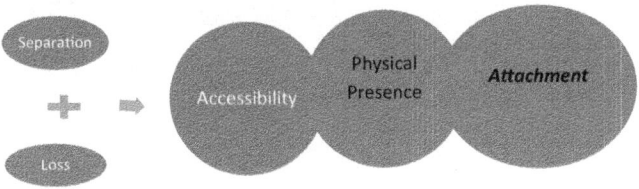

Fig. 6.1 Basic principles of attachment theory

continuity of communication in their relationship despite occasional separations. Indeed research has shown that separation anxiety declines by the third and fourth years of a child's life (Kobak and Marsden 2008); see for example Fig. 6.2.

When children are securely attached, they are better able to let go of their mother and explore their surroundings. An insecurely attached infant or child will be distressed and unable to tolerate the distance between herself and her mother. The reasoning behind this is because a securely attached infant or child has experience of open lines of communication between herself and her mother, and she knows the mother will respond to communications from herself if she needs that from the mother. She/he also knows the mother will comfort her again when they reunite.

Maternal sensitivity to the infant or child's communications may also be affected by their separations. Even if a mother is present in the infant or child's life in terms of proximity, if she lacks understanding or sensitivity to the child's communications, or if she is emotionally unresponsive, there can be effects upon the parent-child attachment relationship. For example, if a mother is not sensitive and emotionally responsive to the infant, the mother-baby attachment may not develop securely (De Wolff and van IJzendoorn 1997). If on the other hand a mother's maternal behaviour can be modified to improve the child's security, her responsiveness to the infant or child can be increased, leading to better securely attached babies and children (see, e.g. Berlin et al. 2008), shown in Fig. 6.3.

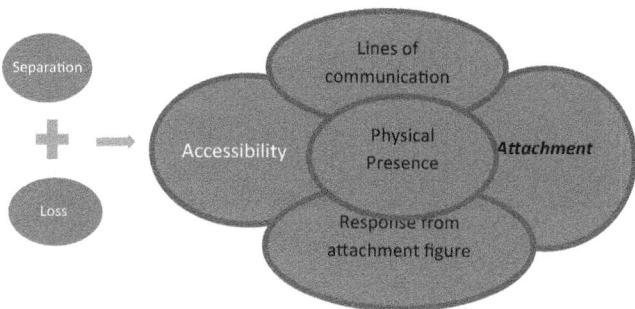

Fig. 6.2 Showing the importance of communication and responsiveness to attachment formation

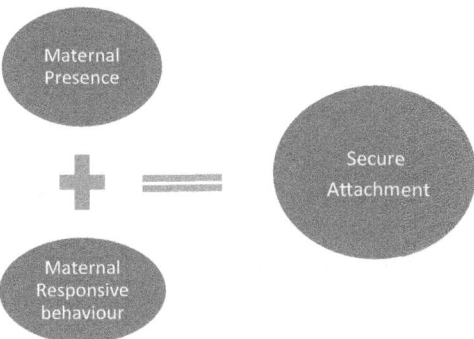

Fig. 6.3 Showing the importance of maternal sensitivity to attachment

Even separations of a baby in daytime childcare can have physiologically stressful effects on a child (Luecken and Lemery 2004). Since instability in a family environment has also been linked with less sensitive and less responsive parenting (Cavanah and Huston 2006) and the effects of even very short-term separations from a mother in the first two years of a baby's life can have lasting behavioural effects, Howard et al. (2011) suggest there is further cause for concern. Finally, parenting following repeated attempts at IVF has also been associated with lowering self-confidence in parenting ability (Torr 2001). Taken together, research has shown that even short periods of separation from a primary caregiver can have severe and lasting effects on secure attachment formations and disruptive behaviours in a child as it grows up.

Attachment styles also influence life satisfaction and later health behaviours. Theories of wellbeing, developed to addresses the importance of different aspects of perceived physical and mental wellbeing, like the subjective wellbeing model for example (Diener and Chan 2011), links emotions, restraint, consumption, income and attachment as significant determinants of wellbeing. Subjective wellbeing and life satisfaction in turn have been linked to higher self-esteem and lower anxiety and depression (Steger et al. 2006) and positive affectivity (King et al. 2006). Regulation of emotions associated with attachments is critical for the development of ill health (DeSteno et al. 2013). For example, relationships between insecure attachment towards the mother as well as the father and dietary and weight problems in children are repeatedly found

(Goossens et al. 2012; Faber and Dube 2015). Over the longer term in addition to the regulation of emotion, parental attachment has also been shown to be critical in the formation of drinking, substance and food consumption habits (see, e.g. Marks 2015). This suggests that attachment opportunities in infancy and childhood may be responsible for lifelong overall wellbeing. It therefore needs to feature prominently in surrogate arrangements policy and practice, since those involved in this process will potentially be responsible for the lifetime wellbeing and life satisfaction experienced by many thousands of people conceived in this way.

Prenatal Attachment

Relationships between the most common primary caregiver—the child-bearing mother—start before birth during the gestation period (Condon 1993; Cranley 1981; Laxton-Kane and Slade 2002). Theoretically, women develop varying degrees of attachment to their foetus during pregnancy and after delivery (Rubin 1984). This relationship or connection between the mother and foetus is composed of emotional and cognitive bonds (Doan and Zimmerman 2003), which according to Condon's (1993) theoretical model includes the presence or absence and strength or weakness of this emotional and cognitive bond. It is important, because better levels of the maternal-foetal relationship are associated with good health behaviours during pregnancy, positive birth outcomes and a higher level of infant development (Alhusen et al. 2013; Lindgren 2001; Walsh et al. 2013) and post-natal maternal involvement (Siddiqui and Haggloff 2000). Women who had a poorer maternal-foetal relationship had more mood instability, depression and irritability towards the baby (Alhusen et al. 2013; Brandon et al. 2009; Pollock and Percy 1999). Recent research on traditional parents pregnant with non-IVF babies has shown that mothers who had a better maternal sensitivity with their baby at six months had previously shown a higher quality of maternal-foetal relationship (Maas et al. 2016). This study therefore shows that the relationship in utero is already predictive of relationships in the early post-natal period. This is of concern to surrogate pregnancies as surrogates try to detach from the foetus, and commissioning parents have no input into these developing pre-birth relationships.

Surrogate and Commissioning Parents' Attachments to the Surrogate Baby

Little research has specifically measured a surrogate mother's attachment to her genetically related (genetic surrogacy) or unrelated (gestational surrogacy) baby. Since prenatal attachment is believed to be influenced by a number of factors including maternal age and attitude towards the pregnancy (Marteau et al. 1988; Siddiqui et al. 1999), a surrogate's ability to relinquish the baby may be possible because surrogate mothers are mostly in their late 20s and 30s and have completed their own family (Blyth 1994; Edelmann et al. 1994). Surrogates are advised by their surrogate agency to ensure they understand the baby is not theirs, so that they do not allow themselves to be attached to the baby or infant following delivery. Handing the baby over to the commissioning couple immediately following the delivery also reinforces the advice. The agencies help surrogates in reconciling their own maternal thoughts and feelings by cognitively restructuring these feelings to match their behaviours (relinquishment of the baby) (van den Akker 2005). van den Akker (2003) found that 15.4% of genetic surrogates and 10% of gestational pregnant surrogates feared they might want to keep the baby. However, surrogate mothers report being less attached to the foetus (Fischer and Gillman 1991; van den Akker 2007) and baby following delivery (van den Akker 2007). In the 2007 study (van den Akker 2007) surrogates who had less positive attitudes towards the foetus were less worried about the health and wellbeing of the baby than commissioning mothers in trimester 1 of pregnancy. At the second trimester, commissioning mothers were significantly more negative about the pregnancy, the foetus and were more worried and more positive about the foetus than surrogate mothers. In the final trimester of pregnancy, commissioning mothers again had more negative attitudes to the surrogate pregnancy and were worried about the health and wellbeing of the baby but their attitudes towards the foetus were significantly more positive than surrogates. Surrogates were therefore characterised by low scores on attitudes towards the foetus during pregnancy, and if they did express their feelings, they were less positive and less concerned about the foetus. Commissioning mothers on the other hand held both positive and negative attitudes towards the pregnancy (once the pregnancy was showing) and some negative attitudes towards the foetus, possibly, because they were significantly more worried about the health and wellbeing of the foetus.

Significant differences between surrogate and commissioning mothers in their attitudes towards the foetus were therefore apparent during the course of the pregnancy. The commissioning mothers' mixed responses show concerns coupled with positive feelings towards the foetus which are likely to reflect an attempt to form a bond or attachment with the foetus (Rubin 1984; Reading et al. 1984). For surrogates, their relatively flat scores on attitudes towards the pregnancy may reflect their attempt to dissociate meaning to the pregnancy to remain detached from it. According to attachment theory, attachment in pregnancy continues to the baby following delivery (Rubin 1984; Reading et al. 1984). Not attaching to the foetus may help them cope and minimise potential feelings of loss at relinquishment. Fischer and Gillman (1991) had already shown that surrogates are less attached to the foetus, confirming Blyth's (1994) retrospective study. van den Akker's (2007) longitudinal study showing that detachment is reported early and maintained throughout the pregnancy with little variation post-delivery suggests these findings are probably robust.

Intended parents do not as a rule experience direct physical contact with the new born baby (Fischer and Gillman 1991) particularly in cross-border surrogacy arrangements, where there may well be a delay of a few days or weeks before they meet for the first time. Intended parents also do not have the hormonal changes accompanying the labour/delivery and responses to the infants crying and needs. Some intended mothers try to breastfeed the surrogate baby. Oxytocin is responsible for, and facilitates, breastfeeding in the mother and intended mothers can be hormonally prepared for breastfeeding. It has been argued that an infant's attachment to the nipple in the first hour following delivery changes the mother's behaviour towards the infant (Widstrom et al. 1990) and allows the infant to recognise its mother's face by increasing its sucking behaviours when shown an image of their mother's as opposed to a stranger's face (Walton et al. 1992). These findings led Sharan et al. (2001) to expose intended mothers to their surrogate-born infant immediately after birth to enhance that process of bonding and recognition. Sharan et al. (2001) reported on two case studies of intended mothers and their genetically related surrogate baby, in a 'rooming-in' hospital setting immediately after birth to encourage the bonding process. The process was successful and mothers were satisfied with the intervention, recognised their babies' needs and temperament and felt more confident taking the baby home two days after the delivery.

The surrogate mothers' own children have not been addressed in this book. However, the unique problems associated with their welfare are recognised. In some cases, the surrogate mother in cross-border surrogate motherhood arrangements may need to leave their own children behind and move into surrogate mother hostels or clinics until birth. In these cases (see, e.g. 'the House of Surrogates' and Chap. 8), the mothers may have to leave their children with their husbands, other family members or neighbours for much of the duration of the pregnancy. These children are at risk of feeling abandoned, and may experience separation anxiety, and may even fear being 'given away' themselves (Steadman and McCloskey 1987). However, this is the topic for another book.

Summary

In summary, separation between a mother/parent and child and attachment formations between them are important to the future health and wellbeing of the child. This chapter has shown that the effects of a surrogate mother's detachment in pregnancy may affect the child later on. It has also shown that a commissioning parents' attachment to the surrogate baby will be critical for the child's health and wellbeing in the short and long term. Parenting and separating from a surrogate baby is therefore complex. Perhaps a viable solution to effective parental relinquishing and bonding with the surrogate baby is to confer parenthood status to all those involved; the donors, surrogate mothers and intended parents to secure the welfare of the surrogate-born child. This way the moral responsibility also lies with all parties; the psychological and social responses of all parties are tailored appropriately and acceptable to all and no one needs to be denied their place in this multiparental process. The legal and contractual parental status can reflect the psychosocial, biological, genetic, moral and ethical solution, allowing the child the true and accurate access to its legal, social, psychological, biological and genetic parents who contributed to his or her existence. Positive relationships between the parties are therefore necessary to make informed decisions on the future welfare of the child. This implies, as van Zyl and van Niekerk (2000) said, that altruistic surrogacy is morally preferable to commercial surrogacy, because the surrogate mothers' responsibilities are not turned on or off with a contract or money, but are based upon a relationship between the different sets of parents to care for the child if the other relinquishes it to their care. Moreover, this does not threaten but enhances the modern family in all its varied forms. The future acceptable face of parenthood is therefore no longer based upon the traditional nuclear family, but is more flexible in its interpretation whilst maintaining responsibility for the conceptions created within them.

References

Ainsworth, M. (1990). Some considerations regarding theory and assessment relevant to attachments beyond infancy. In M. T. Greenberg, D. Cicchetti, & E. M. Cummings (Eds.), *Attachment in the preschool years: Theory, research and intervention* (pp. 463–488). Chicago: University of Chicago Press.

Alhusen, J. L., Hyat, M. J., & Gross, D. (2013). A longitudinal study of maternal attachment and infant developmental outcomes. *Archives of Women's Mental Health, 16*, 521–529.

Baslington, H. (2002). The social organization of surrogacy: Relinquishing a baby and the role of payment in the psychological detachment process. *Journal of Health Psychology, 7*(1), 51–71.

Belsky, J. (2005). The developmental and evolutionary psychology of intergenerational transmission of attachment. In C. S. Carter, L. Ahnert, K. Gtrossmann, S. Hrdy, M. Lamb, S. Porges, & N. Sachser (Eds.), *Attachment and bonding: A new synthesis* (pp. 169–198). Cambridge, MA: MIT Press.

Berlin, L. J., Zeanah, C. H., & Luieberman, A. (2008). Prevention and intervention programs for supporting early attachment security. In J. Cassidy & P. R. Shaver (Eds.), *Handbook of attachment* (Vol. 2, pp. 745–761). New York: Guilford Press.

Blyth, E. (1994). "I wanted to be interesting, I wanted to be able to say 'I've done something with my life'": Interviews with surrogate mothers in Britain. *Journal of Reproductive and Infant Psychology, 12*, 189–198.

Bowlby, J (1969/1982) *Attachment and loss: Vol I: Attachment.* New York: Basic Books.

Bowlby, J. (1973). *Attachment and loss: Vol II: Separation.* New York: Basic Books.

Bowlby, J. (1988). *A secure base: Clinical applications of attachment theory* (Vol. 393). London: Routledge.

Brandon, A., Pitts, S., Denton, W., Stringer, A., & Evans, H. (2009). A history of the theory of prenatal attachment. *Journal of Prenatal and Perinatal Psychology and Health, 23,* 201–222.

Bronte-Tinklew, J., Carrano, J., Horowitz, A., & Kinukawa, A. (2008). Involvement among resident fathers and links to infant cognitive outcomes. *Journal of Family Issues, 29,* 1211–1244.

Cavanah, S., & Huston, A. (2006). Family instability and children's early problem behaviour. *Social Forces, 85,* 551–581.

Colpin, H. (2002). Parenting and psychosocial development of IVF children. Review of the research literature. *Developmental Review, 22*(4), 644–673.

Condon, J. T. (1993). The assessment of antenatal emotional attachment: Development of a questionnaire instrument. *British Journal of Medical Psychology, 66,* 167–183.

Cranley, M. S. (1981). Development of a tool for the measurement of maternal attachment during pregnancy. *Nursing Research, 30,* 281–284.

Crawford, T., Cohen, P., Chen, H., Anglin, D., & Ehrensaft, M. (2009). Early maternal separation and the trajectory of borderline personality disorder symptoms. *Development and Psychopathology, 21,* 1013–1030.

De Carli, P., Tagini, A., Sarracino, D., Santona, A., & Parolin, L. (2015). Implicit attitude toward caregiving: The moderating role of attachment styles. *Frontiers in Psychology, 6,* 1–11.

De Wolff, M., & van IJzendoorn, M. (1997). Sensitivity and attachment: A meta-analysis on parental antecedents of infant attachment. *Child Development, 68,* 571–591.

DeSteno, D., Gross, J., & Kubzansky, L. (2013). Affective science and health: The importance of emotion and emotion regulation. *Health Psychology, 32*(5), 474–486.

Diener, E., & Chan, M. Y. (2011). Happy people live longer: Subjective wellbeing contributes to health and longevity. *Applied Psychology: Health and Well-being, 3,* 1–43.

Doan, H. M., & Zimmerman, A. (2003). Conceptualizing prenatal attachment: Toward a multidimensional view. *Journal of Prenatal and Perinatal Psychology and Health, 18,* 109–129.

Edelmann, R., Humphrey, M., & Owens, D. J. (1994). The meaning of parenthood and couples' reactions to male infertility. *British Journal of Medical Psychology, 67*(3), 291–299.

Faber, A., & Dube, L. (2015). Parental attachment insecurity predicts child and adult high caloric food consumption. *Journal of Health Psychology, 20*(5), 511–524.

Fischer, S., & Gillman, L. (1991). Surrogate motherhood: Attachment, attitudes and social support. *Psychiatry, 54*, 13–20.

Gibson, F. L., Ungerer, J. A., McMahon, C., et al. (2000a). The mother–child relationship following in vitro fertilisation (IVF): Infant attachment, responsivity and maternal sensitivity. *Journal of Child Psychology & Psychiatry, 41*, 1015–1023.

Golombok, S., & Tasker, F. (1997). Do parents influence the sexual orientation of their children? Findings from a longitudinal study of lesbian families. *Developmental Psychology, 32*, 3–11.

Golombok, S., Spencer, A., & Rutter, M. (1983). Children in lesbian and single parent households: Psychosexual and psychiatric appraisal. *Journal of Child Psychology and Psychiatry, 24*, 551–572.

Golombok, S., Cook, R., Bish, A., et al. (1995). Families created by the new reproductive technologies: Quality of parenting and emotional development of the children. *Child Development, 66*, 285–298.

Golombok, S., Brewaeys, A., Cook, R., et al. (1996). The European study of assisted reproduction families: Family functioning and child development. *Human Reproduction, 11*, 2324–2331.

Golombok, S., MacCallum, F., & Goodman, E. (2001). The 'test-tube' generation: Parent–child relationships and the psychological well-being of in vitro fertilization children at adolescence. *Child Development, 72*, 599–608.

Golombok, S., Blake, L., Slutsky, J., et al. (2017). Parenting and the adjustment of children born to gay fathers through surrogacy. *Child Development.* doi:10.1111/cdev.12728.

Goossens, L., Braet, C., van Durme, K., Decaluwe, V., & Bosmans, G. (2012). The parent-child relationship as predictor of eating pathology and weight gain in pre-adolescents. *Journal of Clinical Child and Adolescent Psychology, 41*, 445–457.

Habib, C., & Lancaster, S. J. (2005). The transition to fatherhood: The level of first-time fathers' involvement and strength of bonding with their infants. *Journal of Family Studies, 11*, 249–266.

Hahn, C.-S., & DiPietro, J. (2001). In vitro fertilization and the family: Quality of parenting, family functioning, and child psychosocial adjustment. *Developmental Psychology, 37*, 37–48.

Horsey, K. (2015). *Surrogacy in the UK: Myth busting and reform.* Report of the Surrogacy UK. Working Group on Surrogacy *Law Reform*, Surrogacy UK, November.

Howard, K., Martin, A., Berlin, L., & Brooks-Gunn, J. (2011). Early mother-child separation, parenting and child well-being in early head-start families. *Attachment and Human Development, 13*(1), 5–26.

Jadva, V., Blake, L., Casey, P., & Golombok, S. (2012). Surrogacy families 10 years on: Relationship with the surrogate, decisions over disclosure and children's understanding of their surrogacy origins. *Human Reproduction, 27*(10), 3008–3014.

Jones, J., Cassidy, J., & Shaver, P. (2015). Parents' self-reported attachment styles: A review of links with parenting behaviors, emotions and cognitions. *Personality & Social Psychology Review, 19*, 44076.

King, L., Hicks, J., Krull, J., & Del Gaiso, A. (2006). Positive affect and the experience of meaning in life. *Journal of Personality and Social Psychology, 90*, 179–196.

Kobak, R., & Marsden, S. D. (2008). The emotional dynamics of disruptions in attachment relationships: Implications for theory, research and clinical intervention. In J. Cassidy & P. R. Shaver (Eds.), *Handbook of attachment* (Vol. 2, pp. 23–47). New York: Guilford Press.

Lawrence, C., Carlson, E. A., & Egeland, B. (2006). The impact of foster care on development. *Development and Psychopathology, 18*, 57–76.

Laxton-Kane, M., & Slade, P. (2002). The role of maternal prenatal attachment in a woman's experience of pregnancy and implications for the process of care. *Journal of Reproductive and Infant Psychology, 20*, 253–266.

Leerkes, E., & Siepak, K. (2006). Attachment linked predictors of women's emotional and cognitive responses to infant distress. *Attachment, Human Behaviour and Development, 8*, 11–32.

Leventhal, T., & Brooks-Gunn, J. (2000). *"Entrances" and "exits" in children's lives: Associations between household events and test scores.* New York: Teachers College, Columbia University. Unpublished manuscript cited in: Howard, K., Martin, A., Berlin, L., & Brooks-Gunn, J. (2011). Early mother-child separation, parenting and child well-being in early head-start families. *Attachment and Human Development, 13*(1), 5–26.

Levy-Shiff, R., Vakil, E., Dimitrovsky, L., et al. (1998). Medical, cognitive, emotional and behavioural outcomes in school-age children conceived by in vitro fertilization. *Journal of Clinical Child Psychology, 27*, 320–329.

Lindgren, K. (2001). Relationships among maternal-fetal attachment, prenatal depression, and health practices in pregnancy. *Research in Nursing and Health, 24*, 203–217.

Luecken, L., & Lemery, K. (2004). Early caregiving and physiological stress response. *Clinical Psychology Review, 24*, 171–191.

Maas, A. J. B. M., de Cock, E. S. A., Vreeswijk, C. M. J. M., Vingerhoets, A. J. J. M., & van Bakel, H. J. N. (2016). A longitudinal study on the maternal-fetal relationship and postnatal maternal sensitivity. *Journal of Reproductive and Infant Psychology, 34*(2), 110–121.

Marks, D. (2015). Homeostatic theory of obesity. *Health Psychology Open, 1*, 1–30.

Marteau, T., Johnson, M., Plenicar, M., Shaw, R., & Slack, J. (1988). Development of a self-administered questionnaire to measure women's knowledge of prenatal screening and diagnostic tests. *Journal of Psychosomatic Research, 32*(4–5), 403–408.

McMahon, C. A., Ungerer, J. A., Tennant, C., et al. (1997). Psychosocial adjustment and the quality of the mother–child relationship at four months postpartum after conception by in vitro fertilization. *Fertility & Sterility, 68*, 492–500.

Mikulincer, M., & Shaver, P. (2012). An attachment perspective on psychopathology. *World Psychiatry, 11*, 11–15.

Minai, J., Suzuki, K., Takeda, Y., Hoshi, K., & Yamagata, Z. (2007). There are gender differences in attitudes toward surrogacy when information on this technique is provided. *European Journal of Obstetrics and Gynecology and Reproductive Biology, 132*, 193–199.

Mooney-Somers, J., & Golombok, S. (2000). Children of lesbian mothers: From the 1970s to the new millennium. *Sexual and Relationship Therapy, 15*(2), 121–126.

Moss, S. Z., & Moss, M. S. (1975). Surrogate mother-child relationships. *American Journal of Orthopsychiatry, 43*(5), 382–390.

Moss, E., Cyr, C., Bureau, J., Tarabulsy, G., & Dubois-Comtois, K. (2005). Stability of attachment during the preschool period. *Developmental Psychology, 41*, 773–783.

Panorama. (2013). House of surrogates. Tuesday 1 October at 21:00 BST, *BBC Four*. http://www.bbc.co.uk/programmes/b03c591s

Pollock, P. H., & Percy, A. (1999). Maternal antenatal attachment style and potential fetal abuse. *Child Abuse and Neglect, 23*, 1345–1357.

Purewal, S., & van den Akker, O. B. A. (2007). The socio-cultural and biological meaning of parenthood. *Journal of Psychosomatic Obstetrics and Gynecology, 28*(3), 79–86.

Ragone, H. (1994). *Surrogate motherhood: Conception in the heart*. Boulder: Westview Press.

Reading, A., Cox, D., Sledmere, S., & Campbell, S. (1984). Psychological changes over the course of pregnancy: A study of attitudes towards the foetus/ neonate. *Health Psychology, 3*, 211–221.

Robertson, J. (1990). Procreative liberty and the state's burden of proof in regulating noncoital reproduction. In L. Gostin (Ed.), *Surrogate motherhood: Politics and privacy* (p. 35). Bloomington/Indianapolis: Indiana State University Press.

Rubin, R. (1984). *Maternal identity and the maternal experience.* New York: Springer.

Sharan, H., Yahav, J., Peleg, D., Ben-Rafael, Z., & Merlob, P. (2001). Hospitalization for early bonding of the genetic mother after a surrogate pregnancy: Report of two cases. *Birth, 4*, 270–273.

Siddiqui, A., & Haggloff, B. (2000). Does maternal prenatal attachment predict postnatal mother-infant interaction? *Early Human Development, 59*, 13–25.

Siddiqui, A., Hagglof, B., & Eisemann, M. (1999). An exploration of prenatal attachment in Swedish expectant women. *Journal of Reproductive and Infant Psychology, 17*(4), 369–380.

Steadman, J., & McCloskey, G. (1987). The prospect of surrogate mothering: Clinical concerns. *Canadian Journal of Psychiatry, 32*(7), 545–550.

Steger, M., Frazier, P., Oishi, S., Kaler, M., et al. (2006). The meaning of life questionnaire: Assessing the presence of and search for meaning in life. *Journal of Counseling Psychology, 53*, 80–93.

Tizard, B., & Hodges, J. (1978). The effects of early institutional rearing on the development of eight year old children. *Journal of Child Psychology and Psychiatry, 19*, 971–975.

Torr, M. (2001). *The experience of pregnancy and parenthood after assisted conception.* London: AceBabes.

van Balen, F. (1996). Child-rearing following in vitro fertilization. *Journal of Child Psychology & Psychiatry, 37*, 687–693.

van den Akker, O. B. A. (2003). Genetic and gestational surrogate mothers' experience of surrogacy. *Journal of Reproductive and Infant Psychology, 21*, 145–161.

van den Akker, O. B. A. (2005). A longitudinal pre pregnancy to post delivery comparison of genetic and gestational surrogate and intended mothers: Confidence gynecology. *Journal of Psychosomatic Obstetrics and Gynecology, 26*(4), 277–284.

van den Akker, O. B. A. (2007). Psychological trait and state characteristics, social support and attitudes to the surrogate pregnancy and baby. *Human Reproduction, 22*(8), 2287–2295.

van den Akker, O. B. A. (2012). *Reproductive health psychology*. Chichester: Wiley-Blackwell. ISBN-13: 978-0470683385.

van den Akker, O. B. A. (2015). Chapter: Emotional and psychosocial risk associated with fertility treatment. In R. Mathur (Ed.), *Reducing risk in fertility treatment*. London: Springer Science + Media.

van Zyl, L., & van Niekerk, A. (2000). Interpretations, perspectives and intentions in surrogate motherhood. *Journal of Medical Ethics, 26*, 404–409.

Walsh, J., Hepper, E. G., Bagge, S. R., Wadephul, F., & Jomeen, J. (2013). Maternal-fetal relationships and psychological health: Emerging research directions. *Journal of Reproductive and Infant Psychology, 31*(5), 490–499.

Walton, G., Bower, N., & Bower, T. (1992). Recognition of familiar faces by newborns. *Infant Behaviour and Development, 15*, 265–269.

Weaver, S. M., Clifford, E., Gordon, A. G., et al. (1993). A follow-up study of 'successful' IVF/GIFT couples: Social-emotional well-being and adjustment to parenthood. *Journal of Psychosomatic Obstetrics and Gynecology, 14*(special issue), 5–16.

Widstrom, A., Wahlberg, V., Matthiesen, A., et al. (1990). Short term effects of early suckling and touch of the nipple on maternal behaviour. *Early Human Development, 21*, 153–163.

Wynter, K., Rowe, H., Tran, T., & Fisher, J. (2016). Factors associated with father to infant attachment at 6 months postpartum: A community-based study in Victoria, Australia. *Journal of Reproductive and Infant Psychology, 34*(2), 185–195.

Zeanah, C., Smyke, A. T., Koga, S. F., Carlson, E., & The Bucharest Early Intervention Project Core Group. (2005). Attachment in institutionalized and community children in Romania. *Child Development, 76*, 1015–1028.

7

Individuals Born from Surrogate Mothers

Surrogate-born children are commissioned for and by couples commissioning them. The process involves the intentional creation of a person using third parties and often involves numerous professionals from across disciplines and offices. Their collective actions therefore require additional responsibilities from the very planning stages of the process taking account of the lifelong welfare of the person so created (van den Akker 2013). The availability of and interest in genetic testing for health and family history reasons have led to increasing numbers of individuals using these services. The consequences include adults who were previously unaware of disparities in their family history are now finding out there are significant anomalies in their makeup. This chapter addresses issues of malpractices and highlights areas of concern for the welfare of the child.

Since the circumstances of surrogate-born offspring are different from children put up for adoption and from traditionally conceived children, a revision of priorities for treating patients in assisted conception to focusing on the resultant children should be paramount. Information about the assisted conception processes involved and implications counselling of the consequences of particularly third-party treatment are necessary to shift the patients' focus from believing they are building a traditional

© The Author(s) 2017
Olga B.A. van den Akker, *Surrogate Motherhood Families*,
DOI 10.1007/978-3-319-60453-4_7

family to accept that their family will be alternative—created with genetic and/or gestational difference (Allan 2012). This, according to the Prevention, Outcomes, Consequences model shown below (Fig. 7.1), requires individual and collective responsibility.

Focusing on the latter two aspects of the model—the outcomes and consequences—relates to the children conceived through assisted conception who may be healthy or affected by, for example, inherited damaged DNA (Bonde et al. 2008), some of which may not be expressed until much later on in the adult (surrogacy conceived) individual. Within the model, attention to the treatment which takes place at one point must therefore also include a consideration of the consequences which may take place at another point in time, possibly much later on in a person's life. The gaps in understanding of the long-term effects of different treatments and gaps in the care given to individuals (particularly within the surrogacy arrangements triad) after the treatment has taken place are a significant source of dissatisfaction (van Empel et al. 2010a, b). A holistic/lifespan perspective of short- and long-term care needs, needs to be implemented at the outset (van den Akker 2013).

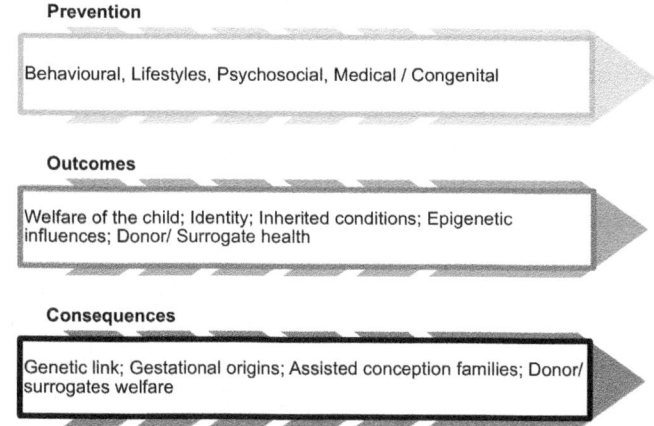

Prevention

Behavioural, Lifestyles, Psychosocial, Medical / Congenital

Outcomes

Welfare of the child; Identity; Inherited conditions; Epigenetic influences; Donor/ Surrogate health

Consequences

Genetic link; Gestational origins; Assisted conception families; Donor/ surrogates welfare

Fig. 7.1 Individual and collective responsibility needs to be taken in research, policy and clinical practice as shown in the Prevention, Outcomes, Consequences (POC) model (Adapted from van den Akker 2013)

The Baby as Outcome

Research, policy and practice into surrogate motherhood usually refer to the outcome in surrogacy as a surrogate baby or child, a convenient and compassionately framed position suggesting that the babies need nurturing and caring parents. However this is a secondary consequence. The commissioning parents' needs are the primary needs being met by surrogacy arrangements and the baby is there to fulfil the parental need. Researchers, policymakers and practitioners therefore need to consider the welfare of the surrogate-born children because their needs are secondary, not primary. One British research group has published most of the follow-up studies of a small number of surrogate-born children using assessments every few years. Children who are no longer interested in participating or are lost to follow-up are replaced by new children, meaning any changes reported with different age groups may be due to other factors and do not represent a full longitudinal perspective. Nevertheless, these studies do provide an insight into surrogate-born children at different ages (Golombok and Tasker 2015). Although the results of psychological assessments are generally acceptable, higher levels of adjustment problems have been reported in surrogate-born children compared to gamete donation children (Golombok et al. 2012), suggesting the effects of surrogacy should not be undermined. However, to date, little information about the needs of surrogate offspring as independent adolescents and none of surrogate-born adults exists. Jadva et al. (2012), from the same research group, reported on 10-year-old adolescents who knew of their surrogate origins that they understood what that meant and the majority of this small sample (13/14) 10-year-olds were in contact with and liked their surrogate mother. It is difficult to think of the psychological adjustment or identity issues experienced in adult surrogate offspring, since nothing is yet known about them. However, what is certain is that these surrogate-born individuals upon reaching adulthood will know (if informed) that they were arranged for the benefit of their parents' needs, usually involving a contract, substantial financial investment and (in genetic surrogacy arrangements) a genetic and gestational link with a surrogate mother who relinquished them as babies.

There is little or no mention either of the family difference that is created—a difference from the traditional biologically and contractually less complicated family. Instead, the parties emphasise that surrogacy is as close to the traditional family as they can get. It is not. There is nothing traditional about using another woman's uterus to bear a foetus and go through the birthing process for the specific purpose to relinquish it upon or soon after delivery, as specified in a contract and brokered via agents. Since a surrogate-born individual is specifically commissioned, it is important to acknowledge the reproductive process is unusual in numerous ways. There is even less commonality with the norm when donor gametes or embryos are used in the surrogate pregnancy, involving a substantial amount of time, effort and expense in the multidisciplinary clinical interventions. Clinical reproductive interventions do not only lead to genetic difference (as in the extreme example of mitochondrial donation, or the more common use of gamete or embryo donation), they also lead to epigenetic difference, the mechanisms that differentiate us in the very early stages of life. The denial of difference has important implications for surrogate conceived adults across the world. Not only is the decision made for them that a surrogate mother (with or without donated gametes or embryos) is not important as a mother, many offspring who are aware of their conception will be wondering about the unknown genetic and/or gestational environment which has contributed to their eventual makeup.

The commercialisation of surrogate motherhood is a transaction both sets of parents are party to. Individuals who have suffered injury or harm through medical negligence (e.g. a suboptimal surrogate mother environment) can claim legal compensation from other people or organisations responsible. Although the compensation culture is not yet embedded in countries across the globe, it is only a matter of time before people will start thinking that children should be able to sue parents for illness caused by passive smoking (Ferriman 1993), drinking excessive alcohol (Linder 2005) or that might have been prevented by vaccination (Rodal and Wilson 2010) or even for wrongful birth (Wilmoth 1980). In the United Kingdom there is no precedent to such cases succeeding, which has been attributed to the 'importance the legal system attaches to the integrity of the family as a social unit' (Larcher and Brierley 2014). However, the conviction of the integrity of the family is put into question in third-party conception and notably in commercial surrogate motherhood.

In practice, there are implications for the adolescents and adults conceived via surrogacy particularly if donor gametes are used. News media regularly feature cases of malpractice, for example when gamete donors are used and not screened properly. A recent case, highlighted by Stapleton (2016) for CNN news, reported the donor was registered as 'handsome and healthy, with several degrees and a genius-level IQ', but the 36 children born from his sperm were in fact the genetic children of a convicted felon and college dropout who was diagnosed as having schizophrenia, posing a genetic risk to the offspring for schizophrenia between 70% and 100%. Research is increasingly focusing upon epigenetics, the developmental flexibility of a developing embryo to its environment (quantity and quality of nutrient availability) and compensatory responses which may interact with differential delivery of the needs for the developing foetus (EpiHealth 2016).

Children Produced for a Fee

Researchers, practitioners and policymakers involved in surrogacy may not all fully grasp the need for the paramountcy of the welfare of the child principle. The focus tends to be on the welfare of the parents who desperately want to build their family. This is where the tireless work on (particularly inter-country) adoptions becomes important. A number of inter-country adoptions take place in circumstances where there have been violations of the rights of the children and adults due to poverty and/or discrimination as outlined in the International Social Services report (ISS 2012). The report notes that although adoption is often used as a means to ameliorate rights deprivations and to implement the best interests of the child, in practice it may be driven by the desire of adults for children, and by financial incentives. Since vulnerable children should not have to wait for poverty, discrimination and so on to be resolved in their home countries therefore obliterating the need for interventions to secure good homes for them, placing them up for adoption may be in their best interests. This, argument according to Smolin (2012), provides sufficient ambiguity in practice as to the proper implementation of the welfare of the child principle. These paradoxes also create 'grey zones' (Smolin 2012, p. 3), where illicit and illegal activity can take place, including children kidnapped and sold for inter-country adoption, fraud and money being used as inducements to obtain relinquishments (in some cases this is via 'surrogacy'), and false documentation being supplied to cover up the illegality of the process. These abusive practices occur widespread, and some 'governments, adoption agencies, and adoptive parents, have powerful incentives to deny or minimize the extent of these illicit practices' (Smolin 2012, p. 3).

The ISS (2012) reports upon child production in a number of different countries including those reported in Box 7.1.

Box 7.1: Showing a Number of Cases of Child Production Across the World

- In Brazil (Time 1986) federal agents discovered a maternity hospital referred to as 'baby farms' and a number of clandestine nurseries in a lawyer's house—including 20 children from newborn to three-year-olds.
- In Honduras, The Christian Science Monitor (1993) reported 'fattening houses' or 'casas de engorde' (which are private homes where children are kept until adoptive parents are found).
- In Greece (The Daily Mail 2006), a baby factory controlled by the Bulgarian and Romanian Mafias functioned and flourished. Here women were 'impregnated by mafia racketeers and then, housed, fed and clothed for the next nine months. After the birth, the birth mother took care of her child for 40 days before the child was adopted by a foreign couple in exchange for a large sum of money'.
- In Nigeria (Ijeoma 2009) the Uzoma Clinic is one of several illegal 'baby farms' where infants are sold to people desperate for children and ready to pay to avoid the red tape of the country's domestic adoption laws.

Although British surrogacy is relatively well regulated by the HFEA, there are many other countries where babies are produced under the auspices of clinics for profit. The question may not be where to draw the line but why is the welfare of the child not at the top of the treatment priorities list. Any individual born through these circumstances for a fee will be affected. The previous chapters have already documented some of the conditions surrogate mothers find themselves in, the deprivation that led them to become surrogate mothers, coercion, lack of understanding and lack of informed consent. These are all issues of moral rights and wrongs which form part of the surrogate motherhood arrangements experience (see further Chaps. 8 and 9) and will impact upon the surrogate-born individual as they grow into adulthood.

The Medical Perspectives

Research into the outcome of genetic and gestational pregnancies generally follows research into IVF outcomes, namely pregnancy and live birth rates. Little research has considered the effects of the route to the pregnancy, the psychological state of the surrogate mother who may be trying not to attach to the foetus and her health behaviours influencing the foetus's epigenetic health and future wellbeing. Details on perinatal data are scarce. Parkinson et al.'s (1999) research provides an exception; they reported the incidence of malformations in gestational surrogate babies to be comparable to that reported for the general population. More recent research suggests infertile couples using IVF or ICSI are at a greater risk of premature birth (Koudstaal et al. 2000), pregnancy complications including low birthweight (Schieve et al. 2002) and a higher rate of malformations (Stromberg et al. 2002; Hansen et al. 2002). It is likely to be parental factors that are responsible for these outcomes, rather than factors related to the techniques used, since the statistics for adverse consequences are lower after surrogate pregnancies, which were also included in Schieve et al.'s (2002) data which use the same techniques as in IVF. However, a recent study reported that oocyte donation is associated with lower foetal birthweight, pregnancy complications and an increased incidence of caesarean sections—without clear clinical indications—in pregnancies brought about with donated oocytes (Savasi et al. 2016). Since gestational surrogates receive embryos with 'donated' (from the commissioning mother or a donor) oocytes, both the surrogate and the baby are likely to be at the same risks as oocyte recipients and their babies in IVF treated cycles.

Although there remain some doubts about the reasons for increased adverse outcomes in IVF/ICSI pregnancies (Yeung et al. 2016), and surrogate-born IVF children have less reported adverse outcomes, children born from all techniques must be monitored and new data must incorporate the longer-term psychological health of these children also. Recent data from three-year-old IVF children not conceived via surrogacy, but using the IVF technique, has shown no differences between ART singletons and naturally conceived singletons, at three years old, but a slightly lower weight in ART twins compared with naturally conceived twins (Yeung et al. 2016). There are, however, some concerns that rapid 'catching up' of weight in underweight infants may predispose them to atherosclerosis (Kleijkers et al. 2014) and cardiovascular risk later in childhood (Kelishadi et al. 2015), suggesting long-term research is necessary.

Little attention has been paid to factors involved in the developing embryo and foetus. Even before the pregnancy, in IVF gestational surrogacy the sensitivity of the embryo to its surroundings including the IVF conditions and culture used has been shown to be important in the early (cleavage stage) embryo developmental stages. Similarly, epigenetic and environmental factors can affect gene activity, and how individuals are programmed at an early embryonic stage to risks for certain diseases. The developmental competence of the oocytes also affects gene expression in the embryo. Finally, the genetics of the oocyte 'mother' and dietary status of the surrogate mother may have an impact upon embryo development and gene expression (Gardner and Lane 2004).

Nothing is known about the foetus' health in a surrogate's uterus, or what her contribution to the epigenetic makeup of the new surrogate baby will be. Epigenetic factors here are used in the context of biology as a model for cell differentiation from their early embryonic state. We do know (according to the Barker hypothesis; Barker et al. 1990) that birth weight is linked to the development of certain serious diseases, such as diabetes in later life, cardiovascular disease and high blood pressure, and that IVF babies have a lower birthweight. No one has yet considered the relevance of the effects of a virus, alcohol or drug consumption in surrogate pregnant women. A surrogate mother carries a foetus she tries not to call her own, one whom she hopes not to bond with to ensure the relinquishment will be easier (van den Akker 2003, 2007). She knows she will not care for the child, and in gestational surrogacy, the child will not carry her genetic makeup. However, the foetus is influenced by much of what she does, what she comes into contact with and what she eats, drinks, breathes in and ingests in other ways. Some of these may have severe and undisputed consequences for the developing foetus (Egliston et al. 2007; Ombelet et al. 2005). In women infected with, for example, the Zika virus, in addition to microcephaly, 29% of scans showed abnormalities in babies in the womb, including growth restrictions (Rasmussen et al. 2016). Foetal alcohol syndrome (Jones and Smith 1973) and smoking (Rogers 2008) or drug use including the use of prescription medications (Lammer et al. 1987) could affect the development of the foetus in utero, particularly in the early period of development (van den Akker 2012). At the individual level the risks may be small, but at societal level this can be substantial (Cordis 2016; Epihealth consortium).

The Resultant Surrogate Offspring

Although research has been carried out on the adults involved in surrogate motherhood arrangements, little is known about the children brought about as a result of surrogacy. There are now at least 20,000 children born through cross-border surrogacy (International Reference Centre for the Rights of Children Deprived of Their Family; Institute of Social Sciences 2012), and the United Kingdom alone saw a sharp rise in annual numbers of PO applications after 2007 of children born as a result of surrogate arrangements in the United Kingdom (Crawshaw et al. 2012, 2013a). These figures are likely to be greater, because not all surrogate and commissioning couples feed back to the agencies if and when they successfully completed these arrangements, and some commissioning parents are known not to register their children for POs. It is also likely that arrangements have taken place outside of any involvement of organisations involved in surrogacy, making accurate documentation impossible. The surrogate children born in the United Kingdom, however, are all mostly still relatively young and therefore unable to voice their opinions on what it is like for them to be a surrogate child, although there is some emerging data on how they feel about their genetic or gestational 'mother' (see, e.g. Jadva et al. 2012). Surrogate-born adults' views will no doubt become public in the next few years, and they should help to provide a framework for future practice.

The Effects of Disclosure

Unlike many other forms of creating families the parents of surrogate children tend to be open about their conception, gestation and genetic origins. This is in part because many surrogate and commissioning couples develop a strong bond or friendship during the arrangements because both physical and emotional changes are relevant to both parties. Many couples also experience strong emotions when the result of a pregnancy test is positive and the experience of birth is shared or when a pregnancy has not been established or results in a miscarriage. These emotions are shared by both parties together under these intimate conditions. They therefore get to know each other well and most keep in contact well after the arrangement has terminated. The child then refers to the surrogate as an aunt or 'other' mum.

Telling a child the truth about its origins can be difficult for the parents. This can be more difficult if it is left until later on in a child's life. Research on adoptees and donor conceived adults has shown that most would like to know about their origins and if possible they would like to get to know their 'genetic' parent(s) (Crawshaw et al. 2016). In surrogacy the same applies. A surrogate mother conception is unusual but not uncommon. There are fears of social stigma on the part of some parents (van den Akker 2000, 2001), and fears of who the true genetic parent(s) of the child are leading to the perception of stigma. Commissioning mothers reported concerns about test tube mix-ups in gestational surrogacy and the possibility that the surrogate husband's sperm has fertilised the surrogate rather than the commissioning husbands' sperm in genetic surrogacy. These fears may deter parents from disclosing their unusual reproduction. Even assuming all fertilisation was carried out as planned, parents may find it difficult to tell a child over a certain age that its mum did not carry her/him, or that in fact she/he is the child born via another mother and the current dad. Finding out later in life that donors or surrogate mothers were used could have devastating consequences as is reported in adult donor conceived offspring's experiences (see, e.g. van den Akker 2006; van den Akker et al. 2015) and adoptees. Future surrogate arrangements should draw on the experiences of now grown-up individuals conceived through other means such as adoption and gamete donation. Their views or needs may not be identical but they are more likely to be comparable. In both adoption (particularly inter-country adoption) and gamete donation, accurate record keeping is considered to be good practice so that when children conceived through gamete donation or placed with another family following an adoption wish to search for birth family or genetically related parents or siblings in adulthood, this can be facilitated (Cheney 2016; van den Akker et al. 2015). Many parents who had their babies adopted, donors (Blyth et al. 2017; Crawshaw et al. 2016) and surrogates (van den Akker 2003) also believe contact or access to information is important.

Gamete or embryo donation is also used in a proportion of surrogate arrangements particularly where same-sex, single men or older heterosexual commissioning parents do not wish to use the surrogate's oocyte(s). When the child's origins are not disclosed, accidental disclosure is not only likely (Indekeu et al. 2013) but increasingly certain (Harper et al. 2016). Sperm donation was first recorded in 1884 (see Brewaeys et al. 2005) and oocyte donation commenced in 1983. Historically, all sperm donation was carried out anonymously with even the parents not knowing who the donor was. Recent practices have been more varied with anonymous and known gamete donations a choice to many prospective parents, including prospective parents commissioning a surrogate baby. Gamete and embryo donors have no legal parenthood status in the United Kingdom and are therefore not named on birth certificates and are not financially liable. They also have no legal obligation to the child or have rights over their upbringing (HFEA 2016). The increasing availability of, and interest in, genetic testing to trace ancestors has also led to an increase in individuals knowing or suspecting they are not genetically related to their parents to use these services (Klotz 2016). DNA tests are sought by adult adoptees, foundling and donor conceived individuals in increasing numbers with varying rates of success. However, matches with half siblings and first or second cousins as well as genetic parents have been reported (Petrone 2015). Unintended disclosure by default can also have difficult consequences for those who believed their 'secret' was safe. Research has previously noted that secrets within families are notoriously 'leaky' (Smart 2011). In addition to the unanticipated and potentially devastating consequences of such disclosure in families, there are also dangers of incestuous relationships (Rawstone 2012; Yoffe 2013). Moreover, anonymous donation of oocytes, sperm, embryos or genetic surrogate mothers is likely to be traceable via DNA testing at some point in their life. This means that organisations promising anonymity must prepare anonymous donors and surrogates that there is no guarantee they will remain anonymous. To date, donor anonymity is still practiced (Kramer 2016) when it clearly is not in the best interests of the child.

The Effects of Openness About Origins

Surrogacy agencies such as COTS report there is no indication that any of the surrogate offspring suffer as a result of the knowledge they had acquired at an early age about their unusual origins. Accounts of commissioning mothers have shown that most intend to tell their child from an early age how they were conceived and carried. Linda Nelson, a commissioning mother of twins, tells her story of openness and honesty towards not only her children but the wider social network. She was not aware of any change in other people's attitudes towards the twins (Nelson 1998). Nelson's children understand the commissioning couple's struggle to bring them into the world. Nelson's case is unusual because she used a surrogate to carry her and her husband's embryo's and donated her own oocytes to another infertile couple. She advocated for honesty about this child's genetic origins, and although the daughter born from Linda's donation now lives abroad, she receives progress reports and photographs of the child and the parenting couple have agreed to be truthful to the child about her genetic mother.

Surrogate mothers too believe it is in the best interests of the child to be open about their conception. Kim Cotton's revelations about her first genetic surrogate baby produced some heart-wrenching truths about the disadvantages of 'closed' or anonymous surrogacy. Kim was a genetic surrogate for an anonymous couple and had no contact with the baby following relinquishment, which had 'barbaric' consequences for her and potentially for the surrogate child if she finds out. Her second surrogacy experience was different. Cotton knew the commissioning mother well before offering herself as her surrogate and they have continuing contact. It is likely that the lack of a genetic link between the second surrogacy arrangement (producing twins) and the first also made a difference to her ability to cope with the relinquishment. Linda Nelson (1998), the commissioning mother (mentioned above), explained that her surrogate twins who know of their conception relate to Kim positively. Both parties are satisfied with the children's behaviour, reactions and understanding and neither the surrogate nor the commissioning mother has any regrets about the arrangement or about the children they jointly brought into this world. The common view in surrogacy is therefore that openness about the child's origins is good for the whole family including the child and the surrogate mother.

Child Development

The psychological development of children born as a result of ART— whether donated or own gametes were used—shows generally favourable results (Golombok and Tasker 2015; Ilioi and Golombok 2015; Jadva et al. 2015), although the children studied to date are still relatively young. Future research is advised to focus on adolescent children and adults conceived through assisted conception and surrogacy. Traditional research on loss of a mother and the child's ability to form relationships with the new parent(s) has shown the effects of the loss can be manifold, as shown in Box 7.2:

Box 7.2: Showing the Effects of Loss Upon a Child

- A child can mourn a loss, or fail to complete the mourning or 'loss' process, which is known to have an effect on their own subsequent parenting behaviours;
- How the child perceives himself or herself is also important since its identity may be threatened by the loss and the need to reinvent himself/herself to be accepted by the new parent(s);
- And a child may perceive divided loyalties between the biological and new parent(s), resulting in loneliness and confusion over its identity continuing into adulthood.

There is some evidence suggesting that contact of birth mother with adoptive family and the adopted child is positive, but there is also some evidence suggesting this can compromise the adoptive mother/adoptive child relationship. The same could apply to surrogate mothers staying or not staying in touch with the surrogate child and the commissioning family. The potential risks include for the surrogate child to find out the birth mother was paid to do this, and for the surrogate's own children that she was paid to relinquish a baby. So far, we do not know if these are likely to surface in the future.

Identity

In adoption (Grotevant and von Korff 2011) and transracial adoption in particular (Sherman 2010), identity development has been the focus for research because during the formative years in adolescence (Alvarado et al. 2014) the search for an identity relating to the adoptive or birth family becomes more pressing. The lessons learned from adoption have led to policies and practices acknowledging the harm that can come from secrecy (Brodzinski 2005). Despite this, no efforts have been made to ensure these same rights of gamete donation and surrogacy apply (Sherman et al. 2016). Some adoptees are known to miss the critical identity markers to help them explore their own identity (Alvarado et al. 2014). In adults conceived via gamete donation research has shown a significantly lower collective identity orientation (which refers to a sense of family belonging) in donor offspring compared to the general population (van den Akker et al. 2015). In this study, donor conceived offspring were registered with UK Donor Link to trace their genetic relatives—both potential parent donors and half siblings. Surrogate-born children, particularly cross-border commercial surrogate offspring, are unlikely to know their birth parent (the commercial surrogate mother) and even less likely to know the identity of the gamete donor(s) if they were used. Accurate birth registrations for all children which includes information on genetic and gestational parents would provide the information they need should they find out about their origins. Neither adopted children nor those born from gamete donation are well served by anonymity or from finding out in later life that they are not (or only partially) genetically related to their parents and extended family (Kenny and Higgins 2014, p. 32). Finding out in later life is therefore known to be detrimental.

Statelessness in International Surrogacy

According to Goris et al. (2009), the law of the soil or of the blood (soli or sanguinis) determines citizenship, the latter being the most common. In national and cross-border surrogacy, citizenship is complicated and a child may find itself legally stateless. For example, in the United Kingdom, Norway and New Zealand, a birth mother is considered the legal mother of the child, even if gestational surrogacy with an embryo entirely genetically linked to the intended parents is used. If she is a British national, the child will have her nationality. If the mother is not married and not British, the child automatically receives the nationality of its birth mother. In India, on the other hand, the genetic mother is considered the mother of the child, not the birth mother by virtue of giving birth. However, where donated gametes (in this example, donated oocytes) are used, the nationality goes to the intended parents by virtue of the father's genetic link, not the oocyte donor. National laws across the many nations participating in cross-border surrogacy have rendered children resulting from cross-border arrangements stateless. A child born in India, for example, but destined for Norway as its Norwegian parents provided the gametes will not be considered an Indian citizen because the child is not genetically related to the Indian surrogate mother. At the same time, the Norwegian authorities will not recognise the child as their citizen because the Norwegian mother did not give birth to the child. Some parents have had to wait a long time to legally bring a stateless child into the country to live with them as a family (Mahapatra 2009; Roy 2010; Kroløkke 2012). This separation from family members could affect attachment formations between key individuals.

Although children going through the adoption process are not the same as surrogate children's journeys to new families, there are some similarities. In both cases there are movements of children from one (set) to another (set) of parents. It is necessary to ensure these children end up in a good place. Within the adoption route the new placement will be the only place where children will be able to grow and flourish. However, there are exceptions, and cases of abuse and child trafficking have been reported. National adoptions and surrogacy arrangements are relatively straight-forward as they are subject to national legislation and recommendations (for more details see Chap. 10). International adoptions and cross-border surrogacy (see Chap. 8) are complicated by different laws and customs. The Hague Conference on Private International Law (HCCH) drafted an agreement to ensure global children's rights-based standards for inter-national adoptions. No such agreement exists for children in cross-border surrogacy cases. However, because there are still some major concerns over some international adoptions, and rising concerns over cross-border surrogacy, the HCIA Commission organised a Forum to discuss both international adoptions and cross-border surrogacy.

'The Welfare of the Child' Principle

According to family law, for children who already exist, the child's welfare is the court's paramount consideration with respect to the child's upbring-ing. Specifically, section 1 of The Children Act (1989) is a checklist that includes the following as shown in Box 7.3.

Box 7.3: Showing the Welfare of the Child—The Children Act Checklist

1. The ascertainable wishes and feelings of the child concerned
2. His physical, emotional and educational needs
3. The likely effect on him of any change in his circumstances
4. His age, sex, background and any characteristics of his which the court considers relevant
5. Any harm which he has suffered or is at risk of suffering
6. How capable each of his parents, and any other person in rela-tion to whom the court considers the question to be relevant, is of meeting his needs

---

Concerns that surrogacy represents the commodification of babies and concerns of what the future holds for a surrogate commissioned person if an arrangement goes wrong have been prominent ab initio of surrogacy (Crawshaw et al. 2013b). However, the depth and explicit concerns of psychological or social issues in the welfare of the child principle are vague in relation to the bringing about of surrogate children, and these are not considered at the time the conception is planned. *When* the welfare principle of a child begins in assisted conception and surrogacy—is different in across jurisdictions (de Lacey et al. 2015):

- In the United Kingdom the HFEA (2008) sates: treatment 'should not be provided unless account has been taken of the welfare of any child'.
- In South Australia (1988) and Victoria (2008) the interests of the child who could be born of the procedure are 'paramount'.
- In Canada (2004) it needs to be 'given priority'.
- In Western Australia (1991) it needs to be given 'proper consideration'.

The terminology is generally vague. What, for example, do 'account has been taken', 'paramountcy' 'given priority' and 'proper consideration' of the welfare of the child principle really mean? It has been argued that the stipulation that the interests of a not yet existing child should be considered at the outset may not be practically applicable (Dickey 2011) and that it may not be intended to be practical but a theoretical presupposition (Blyth and Cameron 1998) or an ideology (Blyth et al. 2008). Others point out that the welfare of the child principle at the start of treatment should involve judgements about the parents' plans to provide a good life for the consequent baby (Archard 2004). However, although there is some opposition to this (Jackson 2002), and considerable concern about the potential 'screening' of parents (HFEA 2005), a principle for the welfare of the child before it is even conceived could have true meaning if reasonable plans to care for and nurture the resultant child are assumed (McMillan 2014). It could safeguard against the points raised in this chapter and criminal intent to broker and traffic children or to create children for the purposes of social gain (see also the chapters presented in Part III).

The Psychosocial Welfare

From a psychosocial and cultural perspective, it is important to consider the effects of the arrangement on the person born from this. It has therefore been argued that thinking about the possible outcomes beyond medical outcomes is the responsible route to take. Children born through surrogacy in its many forms may know or find out a number of factors relevant to surrogate motherhood arrangements, including the examples shown in Box 7.4.

Box 7.4: Showing a Number of Factors Relevant in Surrogate Motherhood Practices

- The commissioning couple was able to pay for a surrogate arrangement whilst other people on lower incomes cannot
- Their commissioning (maybe genetic) parent could not carry a pregnancy to term herself
- She had left childbearing too late to have a traditional pregnancy
- The commissioning couple fall out/divorce and one or both no longer want the baby
- Untraceable anonymous gametes/embryos were used in its conception
- That they were conceived via another woman who they may be genetically related to
- It may be a woman they will never know
- She may have been paid
- And may live in another country in poverty
- Or live in relative affluence as a result of being a (repeat) surrogate
- The surrogate saw the surrogate pregnancy and baby different from her non-surrogate pregnancies and children (especially if genetic surrogacy was used)
- The surrogate did not attach to the foetus/baby in pregnancy
- The arrangement was carried out within the family

Surrogacy within the family where the surrogate mother is the sister of the commissioning mother or her own mother means that the child is born via her/his aunt or grandmother. These grandmothers tend to be much older (as in the case of the 58-year-old Ann Stolper who delivered twin granddaughters and 56-year-old Jaci Dalenberg who delivered triplet grandchildren, also for her infertile daughter). A 61-year-old Japanese grandmother gave birth to one grandchild, and the epigenetic environment such considerably older surrogate grandmothers provide for the developing embryo and growing foetus needs to be considered. Because such arrangements are intra-familial they may be a subject to additional epigenetic and social disadvantages or stigma. A person born through surrogacy may have feelings of responsibility about being the cause of potential adverse health consequences or other hardship to the surrogate mother. If the surrogate mother was poor, possibly coerced into an arrangement, illiterate, left her own children for the duration of her surrogate pregnancy, and who may not have given fully informed consent, the surrogate-born offspring may carry a heavy burden. Alternatively, a surrogate mother who claims not to have had any feelings for the pregnancy that gestated the surrogate-born person and had no difficulty giving him/her up for a fee may have effects or consequences upon the surrogate-born child which are relatively unexplored. If these unusually reproduced individuals had automatic access to accurate records detailing the genetic, gestational and social parents, their existence would become understandable without shame. The mysterious circumstances and gaps in identity and health information would be replaced with well-documented facts of life devoid of potentially exploited third parties or of deception.

Summary

Most children born through national surrogate arrangements fare well and are well cared for with a full understanding of their conception, their other 'mother' and why they were conceived in this way. Despite this, genetic, epigenetic, gestational, nutritional, sociopsychological and cultural, ethical, moral effects of a surrogate arrangement on the resultant offspring as noted in this chapter remain relatively unexplored in research, theory, policy and clinical practice. The costs and benefits need to be weighed up against the potential or actual consequences and reflect the short- and long-term welfare of the people emanating from these arrangements, not predominantly the needs of those seeking a surrogate arrangement, as was evident in this chapter. Recommendations for a future policy and practice refresh is necessary worldwide, questioning the ad hoc and ex post facto focus currently adopted in many countries. This should be replaced with a more intense a priori focused medico-legal system drawing more holistically on the medico-legal and sociopsychological aspects of surrogate arrangements on the persons derived therefrom.

References

Allan, S. (2012). Guest Editorial: Donor conception, secrecy and the search for information. *Journal of Medicine and the Law, 19*(4), 631–650.

Alvarado, S., Rho, J., & Lambert, S. (2014). Counselling families with emerging adult transracial and international adoptees. *The Family Journal, 22*(4), 402 408.

Archard, D. (2004). Wrongful life. *Philosophy, 78*, 403–419.

Barker, D., Bull, A., Osmond, C., & Simmonds, S. J. (1990). Foetal and placental size and risk of hypertension in adult life. *British Medical Journal, 301*, 250–262.

Blyth, E., & Cameron, C. (1998). The welfare of the child: An emerging issue in the regulation of assisted conception. *Human Reproduction, 13*, 2339–2355.

Blyth, E., Burr, V., & Farrand, A. (2008). Welfare of the child assessments in assisted conception: A social constructionist perspective. *Journal of Reproductive and Infant Psychology, 26*, 31–43.

Blyth, E., Frith, L., Crawshaw, M., & van den Akker, O. (2017). Gamete donors' motivations for, expectations and experiences of, registration with UK donorlink. *Human Fertility*, 1–11. http://dx.doi.org/10.1080/1464727 3.2017.1292005.

Bonde, J. P., Toft, G., Rylander, L., et al. (2008). Fertility and markers of male reproductive function in Inuit and European populations spanning large contrasts in blood levels of persistent organochlorines. *Environmental Health Perspectives, 116*(3), 269–277.

Brewaeys, A., de Bruyn, J. K., Louwe, L. A., & Helmerhorst, F. M. (2005). Anonymous or identity registered sperm donors? A study of Dutch recipients' choices. *Human Reproduction, 20*, 820–824.

Brodzinski, D. (2005). *Reconceptualizing openness in adoption: Implications for theory, research and practice* (pp. 145–166). New York: Praeger Publishers.

Cheney, K. (2016). Preventing exploitation, promoting equity: Findings from the International Forum on Intercountry Adoption and Global Surrogacy 2014. *Adoption & Fostering, 40*(1), 6–19.

Cordis. (2016). *Maternal womb conditions impact health.* http://cordis.europa. eu/result/rcn/92607_en.html. Accessed 8 May 2016.

Crawshaw, M., Blyth, E., & van den Akker, O. (2012). The changing profile of surrogacy in the UK – Implications for national and international policy and practice. *Journal of Social Welfare and Family Law, 34*(3), 265–275.

Crawshaw, M., Blyth, E., & van den Akker, O. (2013a). The changing profile of surrogacy in the UK: Implications for national and international policy and practice. *Journal of Social Welfare and Family Law, 34*(3), 1–11.

Crawshaw, M., Blyth, E., & van den Akker, O. (2013b). The ethics and aesthetics of paid surrogacy. *Obstetrics, Gynaecology and Midwifery News, April.* 20–23.

Crawshaw, M., Frith, L., van den Akker, O., & Blyth, E. (2016). Voluntary DNA-based information exchange and contact services following donor conception: An analysis of service users' needs. *New Genetics and Society, 35*(4), 372–392.

De Lacey, S., Peterson, K., & McMillan, J. (2015). Child interests in assisted reproductive technology: How is the welfare principle applied in practice? *Human Reproduction, 30*(3), 616–624.

Dickey, A. (2011). The best interests principle: Truth, ideology or mantra? *Australian Law Journal, 85,* 159–164.

Egliston, K. A., McMahon, C., & Austin, M. (2007). Stress in pregnancy and infant HPA axis function: Conceptual and methodological issues relating to the use of salivary cortisol as an outcome measure. *Psychoneuroendocrinology, 32,* 1–13.

EpiHealth. (2016). http://www.epihealthnet.org/. Accessed 12 May 2016.

Ferriman, A. (1993). Children seek to sue parents over passive smoking. The independent. http://www.independent.co.uk/news/children-seek-to-sue-parents-over-passive-smoking-1481858.html. Accessed 3 May 2016.

Gardner, D. K., & Lane, M. (2004). Ex vivo early embryo development and effects on gene expression and imprinting. *Reproduction, Fertility and Development, 17*(3), 361–370.

Golombok, S., & Tasker, F. (2015). Socio-emotional development in changing families. In M. E. Lamb & R. M. Lerner (Eds.), *Handbook of child psychology and developmental science, Social, emotional and personality development* (Vol. 3, 7th ed., pp. 419–463). Hoboken: Wiley.

Golombok, S., Blake, L., Casel, P., Roman, G., & Jadva, V. (2012). Children born through reproductive donation: A longitudinal study of psychological adjustment. *Journal of Child Psychology and Psychiatry, 54*(6), 653–660.

Goris, I., Harrington, J., & Köhn, S. (2009). Statelessness: What is it and why it matters. *Forced Migration Review, 32*(6), 4–6.

Grotevant, H., & von Korff, L. (2011). Adoptive identity. In S. Schwartz, K. Luyckx, & V. Vignoles (Eds.), *Handbook of identity, theory and research* (pp. 585–601). New York: Springer.

Hansen, M., Kurinczuk, J. J., Bower, C., & Webb, S. (2002). The risk of major birth defects after intracytoplasmic sperm injection and in vitro fertilization. *New England Journal of Medicine, 346*, 725–730.

Harper, J. C., Kennett, D., & Reisel, D. (2016). The end of donor anonymity: How genetic testing is likely to drive anonymous gamete donation out of business. *Human Reproduction, 31*(6), 1135–1140.

HFEA. (2005). The Human Fertilisation and Embryology Authority tomorrow's children: Report of the policy review of welfare of the child assessments in licensed assisted conception clinics, London.

HFEA. (2008). Human Fertilisation and Embryology Act. http://Legislation.gov.uk/ukpga/2008/22/contents. Accessed 25 May 2016.

HFEA. (2016). The Human Fertilisation and Embryology Authority. http://www.hfea.gov.uk/donor-conception-births.html. Accessed 2 May 2016.

Ijeoma, E. (2009). Nigerian police crack illicit baby trafficking ring. http://www.reuters.com/article/idUSL12808424

Ilioi, E. C., & Golombok, S. (2015). Psychological adjustment in adolescents conceived by assisted reproductive techniques: A systematic review. *Human Reproduction, 21*(1), 84–96.

Indekeu, A., Dierickx, K., Schotsmans, P., Danielsm, K. R., Rober, P., & d'Hooghe, T. (2013). Factors contributing to parental decision making in disclosing donor conception: A systematic review. *Human Reproduction Update, 19*, 714–733.

International Reference Centre for the Rights of Children Deprived of Their Family; Institute of Social Sciences. (2012). https://www.iss.nl/fileadmin/ASSETS/iss/Guests/Adoption___surrogacy/Publications/Herve_Boechat_Pub.pdf. Accessed 21 Mar 2017.

Jackson, E. (2002). Conception and the irrelevance of the welfare principle. *The Modern Law Review, 2*, 176–203.

Jadva, V., Blake, L., Casey, P., & Golombok, S. (2012). Surrogacy families 10 years on: Relationship with the surrogate, decisions over disclosure and children's understanding of their surrogacy origins. *Human Reproduction, 27*(10), 3008–3014.

Jadva, V., Imrie, S., & Golombok, S. (2015). Surrogate mothers 10 years on: A longitudinal study of psychological wellbeing and relationships with the parents and child. *Human Reproduction, 30*(2), 373–379.

Jones, K. L., & Smith, D. W. (1973). Recognition of the fetal alcohol syndrome in early infancy. *Lancet, 302*, 999–1001.

Kelishadi, R., Haghdoost, A. A., Jamshidi, F., Aliramezany, M., & Moosazadeh, M. (2015). Low birthweight catch-up growth: Which is more associated with cardiovascular disease and its risk factors in later life? A systematic review and cryptanalysis. *Paediatric International Child Health, 35*, 110–123.

Kenny, P., & Higgins, D. (2014). Past adoption practices: Key messages for service delivery responses and current policies. In A. Hayes & D. Higgins (Eds.), *Families, policy and the law: Selected essays on contemporary issues for Australia* (p. 32). Melbourne: Australian Institute of Family Studies.

Kleijkers, S. H., van Montfoort, A. P., Smits, L. J., Viechtbauer, W., Rosenboom, T. J., Nelissen, E. C., Coonen, E., et al. (2014). IVF culture medium affects post-natal weight in humans during the first 2 years of life. *Human Reproduction, 29*, 661–669.

Klotz, M. (2016). Wayward relations – Novel searches of the donor-conceived for genetic kinship. *Medical Anthropology, 35*, 45–57.

Koudstaal, J., Braat, D. D., Bruinse, H. W., et al. (2000). Obstetric outcome of singleton pregnancies after IVF: A matched control study in four Dutch university hospitals. *Human Reproduction, 15*, 1819–1825.

Kramer, W. (2016). DNA donors = not anonymous. https://www.donorsiblingregistry.com/dsr-support-and-info/dna-testing. Accessed 8 May 2016.

Kroløkke, C. (2012). From India with love: Troublesome citizens of fertility travel. *Cultural Politics, 8*(2), 307–325.

Lammer, E. J., Sever, L. E., & Oakley, G. P. (1987). Teratogen update: Valproic acid. *Teratology, 35*(3), 465–473.

Larcher, V., & Brierley, J. (2014). Fetal alcohol syndrome (FAS) and fetal alcohol spectrum disorder (FASD) – Diagnosis and moral policing; an ethical dilemma for paediatricians. *Archives of Diseases in Childhood.* doi:10.1136/archdischild-2014-306774.

Linder, E. (2005). Punishing prenatal alcohol abuse: The problems inherent in utilizing civil commitment to address addiction. *University of Illinois Law Review, 3*, 873.

Mahapatra, D. (2009, December 16). German couple's surrogate kids may end up stateless. *The Times of India.* http://timesofindia.indiatimes.com/india/German-couples-surrogate-kids-may-end-up-stateless/articleshow/3540835.cms

McMillan, J. (2014). Making sense of child welfare when regulating human reproductive technologies. *Journal of Bioethics Enquiry, 11*, 47–55.

Nelson, L. (1998). In E. Blyth, M. Crawshaw, & J. Spiers (Eds.), *Truth and the child 10 years on: Information exchange in donor assisted conception*. Birmingham: British Association of Social Workers.

Ombelet, W., De Sutter, P., Van der Elst, J., & Martens, G. (2005). Multiple gestation and infertility treatment: Registration, reflection and reaction – The Belgian project. *Human Reproduction Update, 11*, 3–14.

Parkinson, J., Tran, C., Tan, T., Nelson, J., & Serafini, P. (1999). Perinatal outcome after in-vitro fertilization surrogacy. *Human Reproduction, 14*, 671–676.

Petrone, J. (2015). As consumer genomics databases swell, more adoptees are finding their biological families. GenomeWeb 2015. http://www.genomeweb.com/applied-markets/consumer-genomics-databases-swell-more-adoptees-are-finding-their-biological. Accessed 2 May 2016.

Rasmussen, S. A., Jamieson, D. J., Honein, M. A., & Petersen, I. R. (2016, April 13). Zika virus and birth defects – Reviewing the evidence for causality. doi:10.1056/NEJMsr1604338.

Rawstone, T. (2012, April 17). The man who fathered 1000 children. *The Mail on Sunday*. http://www.dailymail.co.uk/news/article-2130814/The-man-who-fathered-1-000-children-Theyre-middle-class-living-Britain-idea-extraordinary-story-surrounding-birth.html. Accessed 2 May 2016.

Rodal, R., & Wilson, K. (2010). Could parents be held liable for not immunizing their children? *McGill Journal of Law Health/Revue de Droit et Santé De McGill, 4*, 39–64.

Rogers, J. M. (2008). Tobacco and pregnancy: Overview of exposures and effects. *Birth Defects Research Embryo Today, 84*, 1–15.

Roy, S. (2010, June 9). French gay dad may lose surrogate kids. *The Times of India*. http://timesofindia.indiatimes.com/india/French-gay-dad-may-lose--surrogate-kids/articleshow/6025936.cms

Savasi, V. M., Mandai, L., Laoreti, A., & Cetin, I. (2016). Maternal and fetal outcomes in oocyte donation pregnancies. *Human Reproduction, 22*(5), 620–633.

Schieve, L. A., Meikle, S. F., Ferre, C. C., et al. (2002). Low and very low birth weight infants conceived with use of assisted reproductive technology. *New England Journal of Medicine, 346*, 731–737.

Sherman, R. (2010). A theoretical look at biculturalism in intercountry adoption. *Journal of Ethnic and Cultural Diversity in Social Work, 19*(2), 127–142.

Sherman, R., Misca, G., Rotabi, K., & Selman, P. (2016). Global commercial surrogacy and international adoption: Parallels and differences. *Adoption & Fostering, 40*(1), 20–35.

Smart, C. (2011). Family secrets and memories. *Sociology, 45*(4), 539–553.

Smolin, D. M. (2012). Foreword to: *Investigating the grey zones of intercountry adoption* (pp. 3–4). International Social Service. ISBN 978-2-8399-1000-2.

Stapleton, A. C. (2016, April 20). Sperm donor lied about criminal and mental health history, lawsuit alleges. CNN updated 1407 GMT (2207 HKT). http://edition.cnn.com/2016/04/19/health/sperm-donor-criminal-mental-health-history/. Accessed 12 May 2016.

Strömberg, B., Dahlquist, G., Erickson, A., et al. (2002). Neurological sequelae in children born after in-vitro fertilisation: A population based study. *Lancet, 359*, 461–465.

The Children Act. (1989). legislation.gov.uk. Accessed 29 May 2016.

The Christian Science Monitor. (1993, May 13). *In Honduras, a black market for babies.* http://www.csmonitor.com/1993/0513/13121.html

The Daily Mail. (2006, December 22). The shocking truth about the baby factories. http://www.dailymail.co.uk/femail/article-424450/The-shocking-truth-baby-factories.html

Time. (1986, August 4). Brazil Baby Farm: A luxury adoption business by Wendy Smith. http://content.time.com/time/magazine/article/0,9171,961895,00.html

van den Akker, O. B. A. (2000). The importance of a genetic link in mothers commissioning a surrogate baby in the UK. *Human Reproduction, 15*(8), 110–117.

van den Akker, O. B. A. (2001). The acceptable face of parenthood: Psychosocial factors of infertility treatment. *Psychology Evolution and Gender, 3*(2), 137–153.

van den Akker, O. B. A. (2003). Genetic and gestational surrogate mothers' experience of surrogacy. *Journal of Reproductive and Infant Psychology, 21*, 145–161.

van den Akker, O. B. A. (2006). A review of gamete donor family constructs: Current research and future directions. *Human Reproduction Update, 12*(2), 91–101.

van den Akker, O. B. A. (2007). Psychological trait and state characteristics, social support and attitudes to the surrogate pregnancy and baby. *Human Reproduction, 22*(8), 2287–2295.

van den Akker, O. (2012). *Reproductive health psychology* (pp. 162–165). Chichester: Wiley.

van den Akker, O. B. A. (2013). For your eyes only: Bio-behavioural and psychosocial research objectives. *Human Fertility, 16*(1), 89–93.

van den Akker, O. B. A., Crawshaw, M. C., Blyth, E. D., & Frith, L. J. (2015). Expectations and experiences of gamete donors and donor-conceived adults searching for genetic relatives using DNA linking through a voluntary register. *Human Reproduction, 30*(1), 111–121.

van Empel, I., Nelen, W., Tepe, E., van Laarhoven, E., Verhaak, C. M., & Kremer, J. A. (2010a). Weaknesses, strengths and needs in fertility care according to patients. *Human Reproduction, 25,* 142–149.

van Empel, I. W. H., Aarts, J. W. M., Cohlen, B. J., Huppelschoten, D. A., Laven, J. S. E., Nelen, W., & Kremer, J. A. (2010b). Measuring patient-centredness, the neglected outcome in fertility care: A random multicentre validation study. *Human Reproduction, 25,* 2516–2526.

Wilmoth, D. D. (1980). Wrongful life and wrongful birth causes of action: Suggestions for a consistent analysis. *Marquette Law Review, 63*(4), 3.

Yeung, E. H., Sundaram, R., Bell, E. M., Drushell, C., Kus, C., Xie, Y., & Buck Louis, G. M. (2016). Infertility treatment and children's longitudinal growth between birth and 3 years of age. *Human Reproduction, 31*(7), 1621–1628.

Yoffe, E. (2013, February 19). (Dear Prudence). My wife is my sister. Slate. http://www.slate.com/articles/life/dear_prudence/2013/02/dear_prudence_my_wife_and_i_came_from_the_same_sperm_donor.html. Accessed 2 May 2016.

Part 3

Cross Border Surrogacy, Ethical, Moral, Human Rights Contexts and Legal Frameworks

This part covers a number of considerations which are additional to the major issues already described in Parts I and II relevant to cross-border surrogate motherhood, including the increased potential and reality of the commodification of children and exploitation of surrogate mothers in Chap. 8. The power imbalance between the surrogate mother and intended commissioning parent(s) is not always in favour of the latter and examples illustrating the power which has been assumed by some surrogates also tell a story of potential exploitation of the commissioning parents. This chapter shows that the welfare of the child in cross-border surrogacy is harder to prioritise and is effectively impossible to guarantee. As before, some comparisons between surrogacy and adoption and gamete donation are made, particularly within the context of the right of the child to know its biological and genetic parents. The additional risks evident in cross-border compared to national surrogacy are highlighted in some countries which initially opened their doors to international commissioning parents and subsequently—usually following major scandals—closed these again. This has resulted in other countries to open their doors to benefit from the shifting opportunities to take a slice of this million-dollar industry. The experience of surrogacy or surrogacy-like opportunities in a number of different countries is therefore also covered in this part.

The costs and benefits of national and cross-national surrogacy are closely linked to ethical and moral issues. The main ethical and moral questions surrounding national and international surrogacy are discussed in Chap. 9, as are the medical and health professionals' responsibilities for future families which these professionals are instrumental in creating. The fact that this practice has raised so many moral and ethical issues is in itself a cause for concern and issues warnings for the future of much of assisted conception. If surrogate motherhood arrangements are here to stay, and the evidence of its increasing popularity seems to indicate this will be so, the accurate monitoring and record keeping of treatments carried out, gametes or embryos used and surrogate mothers involved in these processes are going to be critical for a morally and ethically sustainable future of third-party reproductive care. Finally, the human rights abuses associated with national and cross-border surrogacy are not always well served by the legal systems in place. Chapter 10 reflects upon the timelines and changes in recommendations and enquiries leading to surrogate motherhood legislation in the United Kingdom as the world's leading example of these developments in legislation. It provides a useful systematic view of how the British government has changed its legislation on surrogate motherhood over a period of four decades using information derived mainly from committees, inquiries and professional bodies' position statements on surrogacy, not research evidence.

The importance of counselling is recognised from the earliest attempts to influence the practice of surrogacy in the 1980s through to today. The fact that that is recognised throughout as important shows that all those involved in the enquiries accepted the importance of psychological factors in surrogate motherhood arrangements. Finally, in this part too, the issues addressed in one chapter are relevant to the same issues discussed in a different context in another chapter, so again there is overlap in the issues addressed. The research, policy and varied practices described in this part are only part of a much more elaborate whole. The reason why this can never be fully comprehensive is because the processes change on a regular basis in line with social, attitudinal, professional and technological shifts in perspectives. Furthermore, continuously developing research evidence for good practice has shifted medical practices in assisted conception, although there are exceptions. For example, Heneghan et al.

(2016) have highlighted clinical practice as increasingly encouraging infertile couples to pay for additional 'add on' treatments even though the evidence for their efficacy is lacking and may be harmful rather than helpful. The paper highlights the problems of fully informed consent which is clearly competing with marketing strategies for (ineffective/harmful) treatments offered for profit. Increasing treatment costs to the individual patients are likely to send them to other countries to compare prices. Shopping around and a lack of full knowledge of the evidence of the efficacy of some of these treatments are marked by additional moral, ethical and legal dilemmas. The vulnerability of hopeful parents-to-be is used by opportunistic individuals to open their doors to lure patients to their businesses, only to close these doors once reports of malpractice or criminal activity shut these avenues down. Overall, the problems identified in this part include a focus on the substantial inequalities based upon socioeconomic status determining who does and who does not have a baby via surrogacy. Future generations may only wonder why criteria for twenty-first-century unusual reproductive practices, with few exceptions, were available to those who could afford it, not to those who might be the best parents.

Reference

Heneghan, C., Spencer, C., Bobrovitz, N., Collins, R., Nunan, D., Plüddemann, A., Gbinigie, O., Onakpoya, I., O'Sullivan, J., Rollinson, A., Tompson, A., Goldacre, B., & Mahtani, K. (2016). Lack of evidence for interventions offered in UK fertility centres. *British Medical Journal, 355*. doi:10.1136/bmj.i6295.

8

Cross-Border Surrogacy

It is not possible to generalise about national and international or cross-border surrogacy as there are vastly different national laws, clinical success rates, population attitudes and ethical principles operating between the participating countries, but there are similarities in the risks identified. In all cases it is unlikely that a surrogate mother will be socioeconomically better off than the commissioning parent(s), but in one country there will be a certain amount of legal involvement and clinical procedures, whereas in another country the law may be non-existent or ill-prepared to deal with surrogacy—and information provision, economic deprivation and educational and gender inequality can be rife. The multiple risks to parents and children identified in cross-border surrogacy are therefore the focus of this chapter. This chapter also considers briefly the varied and multiple reasons why very different people access cross-border surrogacy.

When national adoptions became more difficult, international adoptions gained in popularity until infertility treatment became more accessible (Sherman et al. 2016). The decline in popularity of adoption was also in part because prospective adoptive parents were less likely to receive healthy babies, as there are more children over the age of 10, and children with special needs needing placements (International Social Service 2014). Although adoption was the main method of family building prior to the

© The Author(s) 2017
Olga B.A. van den Akker, *Surrogate Motherhood Families*,
DOI 10.1007/978-3-319-60453-4_8

upsurge in relatively accessible infertility treatment, increasing numbers of people have turned to commissioning a surrogate baby which can be legally easier and less costly if commissioned overseas. It is also an increasingly popular option to build a family because of the strength of desire for a genetic link with the child. The option of a newborn baby which may be genetically linked to one or both of the parents has created the demand for fertility treatment, leaving adoption behind as a non-preferred option (van den Akker 2001, 2007). Cross-border commercial surrogacy is now also potentially quicker and cheaper than international adoption (Sherman et al. 2016), fuelling this shift in family building focus still further. National surrogacy was popular towards the latter decades of the previous century (van den Akker 2007). National surrogacy such as altruistic surrogacy in the United Kingdom and commercial surrogacy in the United States is more restrictive than surrogacy in developing countries and frequently more expensive. These restrictions and expenses have increasingly shifted the demand for surrogacy to other countries since the turn of the century. The increase in demand for cross-border surrogacy has been reported across the world, including Australia (Cuthbert and Fronek 2013) and the United Kingdom (Crawshaw et al. 2013a). However, this too is not without its problems as is amply discussed in the international media and research reports (e.g. Fixmer-Oraiz 2013; Blyth et al. 2014; Crawshaw et al. 2013b; Cuthbert and Fronek 2013) and see also Chaps. 2 and 6.

Cross-border surrogacy arrangements form part of a multi-million-dollar business, one which many developing countries are keen to have a stake in. The rapidly developing market in surrogacy is of concern because there are no international regulations or guidelines. Even within countries, national regulation may be minimal, non-existent or developed in haste in reaction to adverse publicity of arrangements which provoked international outrage. The practice is riddled with uncertainty about its morality, commodification, parentage, citizenship, attachment and inequalities. For example, in India, surrogacy has been marketed prolifically over a period of many years (2002–2015; Hochschild 2009; Carney 2010). India's reproductive tourism market has come under substantial scrutiny (Pande 2010; SAMA 2012) and in 2013 it passed its first surrogacy regulation incredulously banning single or gay foreign commissioning parents (Bromfield and Smith-Rotabi 2014). In 2015,

India drafted an 'Assisted Reproductive Technology' bill that intended to ban all cross-border surrogacy (see, e.g. Ahmad 2015) because it was obvious that the women were so poor and desperate, they would have done anything for money (according to a representative of the Indian Council for Medical Research, as cited in Ahmad 2015). Recently, these legal changes have led to the shutting down of existing gateways for heterosexual couples too, probably in response to worldwide criticism of the economic and social inequity and the potential exploitation and infringements of human rights of Indian surrogates compared to their wealthier commissioning counterparts (Symons 2015). Now only married Indian couples can use the surrogacy services in India and clinics are banned from 'entertaining any foreigners'. Thailand was subject to a number of specifically abhorrent cases of malpractice, including one where a Western couple refused to take one of a set of twins with Down's syndrome but took the healthy infant. Since then, Thailand too closed its doors to international clients commissioning surrogates in their countries.

However, where one door closes another opens, and 14 clinics are reported to have opened up in Phnom Penh in Cambodia, following the closures of previously favourite destination routes (Cook 2015). Cook describes the surrogacy business in Cambodia as 'murky and corrupt'; it has little experience of surrogacy with its first reported case as recent as 2014, with no legal precedent, and if and when the authorities do focus on this, it is possible it will be treated as human trafficking which will have dire consequences to those involved. This is a risk the intended couples may or may not be aware of, since it is reported that couples do not only choose low-resource countries for a cheaper arrangement, they also choose them because legislation is not restrictive (Rimm 2009). Cambodia also has now closed its doors to international clients seeking surrogate motherhood arrangements.

Why some countries ban all surrogacy arrangements, others ban some types of arrangements and yet others allow some or all types of surrogacy (e.g. some states in America) depend upon ethical and moral viewpoints. Even within Europe there are countries who have legalised altruistic surrogacy (e.g. the United Kingdom), and countries where this is banned altogether (e.g. Italy, Germany and France). Further afield, in Canada, Australia and New Zealand restrictions to surrogacy are also in place,

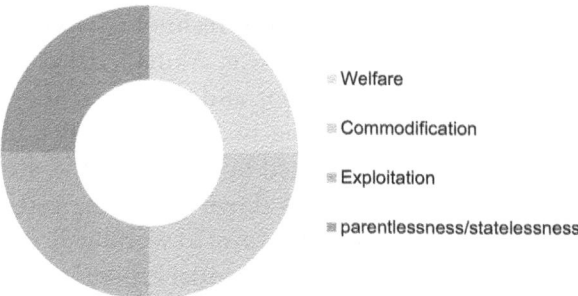

Fig. 8.1 Problems identified in cross-border surrogacy using surrogates from developing nations

whereas Japan and China have bans on all surrogacy. Since there is no overriding regulation able to account for the diverse parenthood and citizen status issues internationally, the legally complicated situation can facilitate exploitation (Rimm 2009). According to Stark (2012) most contemporary international surrogacy arrangements are gestational arrangements.

The main areas of concern in relation to international surrogacy are shown in Fig. 8.1 (adapted from Bromfield and Smith-Rotabi 2014) and are briefly discussed below:

Welfare

Surrogates recruited by some Indian clinics such as the well-known Anand clinic (see BBC 4; 'House of Surrogates' 2013) live in designated dormitories away from their own husband (to avoid infections) and from their children—they spend time being 'farmed'. The positive effects on the surrogate are said to be that they are less likely to be stigmatised by their communities and they receive some skills training, but the negative aspects can be much longer term. For example, the welfare of their own relationship with their husbands could suffer. More worrying is their nearly year-long absences as a mother to their own children. Although no research has yet determined the effects of these separations on the surrogate mothers and her children, psychological research has shown extended separations of a month or more in under five-year-olds can have devastating and long-term consequences including symptoms of borderline personality disorder in adolescence and early adulthood have been reported as a result of such periods of a mother's absence from her young child (Crawford et al. 2009). According to attachment theory, caregivers must be present and accessible if their children are to become attached to them. Bowlby studied children who were separated from their mothers (Kobak and Madsen 2008) and found that separations as short as a week could negatively impact the relationship between mother and child (Bowlby 1969/1982). In addition to Bowlby's pioneering work which focused on the mother's physical presence, Ainsworth (1990) added two other aspects relating to availability that are important for infant attachment: a child's belief that the lines of communication with the mother are open, and the child's belief that the mother will respond if the child needs her. These securities are therefore compromised in the surrogate's own, often young, child(ren)—a fact that has received far too little scientific attention (see Chap. 6).

There is currently also a complete lack of information on the welfare of the children born from cross-border surrogacy arrangements. Their commissioning family is likely to have sufficient income to have initiated the arrangement, and may have shopped around for 'best' possible deals. The surrogate mother is likely to have been extremely poor, uneducated, potentially exploited and may have suffered medically, socially or psychologically as a result of their involvement. It is possible that knowing their birth mother was kept in 'dormitory' or 'farm like' conditions for the duration of the surrogate pregnancy, at the expense of her own family life, may have an effect on them. A child born through such an exploitative arrangement who learns of the surrogate mothers' plight later on in life may feel heartbroken at having had a part to play in that, albeit, one for which they did not consent. Similar feelings of confusion, pity, guilt or anger may become evident in the surrogates' own children.

Commodification

Developing an understanding of the way 'surrogacy deals' are offered and chosen by the commissioning parents as well as the amounts of money and legal paperwork that was involved in their acquisition may make the surrogate-born child/individual feel they were a commodity. Medical 'tourism' which unhelpfully refers to people in one country travelling overseas to obtain health care (such as treatment for disease) or health enhancement services (such as new teeth or other cosmetic surgery) is common and provides cheap alternatives to national options. It may also provide a way around national laws prohibiting commercial surrogacy. Countries market such medical health care and enhancement services as business enterprises, offering free quotations and special 'package' deals to ease the costs (see, e.g. Kohli 2011). Some surrogate industries also offer a holiday attached to the treatment or put clients up in luxury hotels to enhance the experience.

The language used to describe surrogacy practices and surrogate mothers and the intermediaries is often poignant and intended to shock, reflecting the real ethical dilemmas surrogacy poses to many people. It also reflects a persistent negativity and commodification. Corea (1985) was one of the early writers voicing her feminist opinions about surrogacy. In her terms, intermediaries are referred to as 'surrogacy entrepreneurs' and surrogate mothers as 'breeders'. The terms 'contracting out' and 'transacting of babies' reflect her stance of the exploitation of women and commodification of babies. Bromfield and Capous-Desyllas (2012) refer to surrogacy as a form of prostitution. Baby selling practices or child trafficking in surrogacy have also been explored in relation to surrogacy, particularly cross-border surrogacy. However, others note that surrogate motherhood can never be considered baby selling because the contracts are arranged prior to conception (Bromfield and Smith-Rotabi 2014).

Exploitation

According to the SAMA-Resource group for Women and Health (2012) there is a danger that commercial surrogates in developing or low-resource countries are more likely to be exploited with many being uneducated, unemployed and earning a fraction of what the clinic charge (Pande 2009). In developed countries such as the United States, reports on the sociodemographics of commercial surrogates are conflicting. In some reports, surrogates are reported to be more likely to have a considerably more favourable educational and employment status acting as surrogates for 'extra' income (Stark 2012, p. 8), because they enjoy pregnancy (Teman 2008), to develop a sense of empowerment and self-worth (see e.g. Ali and Kelly 2008). They do not focus on the financial gain, but instead emphasise their altruism (Ciccarelli and Beckman 2005) as is reported for altruistic surrogacy in the United Kingdom (Blyth 1994; van den Akker 2003, 2007). However, other research suggests this is an incomplete picture of American surrogates, with many of them going into surrogacy to make up for meagre family incomes thereby supplementing poverty through surrogacy, but without adequate understandings of surrogacy and—if they are already vulnerable—of being exploited (Gugucheva 2010).

America also advertises commercial surrogacy services which offer additional 'perks'. For example, wealthy Chinese couples are keen to pursue surrogacy in the United States, rather than, for example, Thailand, because an American-born surrogate child can obtain US citizenship status by virtue of its birth there. It is also the case that some choose US surrogacy to circumvent their own (now somewhat relaxed) one-child policy, to have taller children (using tall, blonde egg donors) and to have a male child (an option available in some US clinics). However, this is an option only for the particularly affluent international commissioning parent, because these arrangements are likely to set them back by thousands of dollars. Three years ago, 'Chinese customers' were reported to pay a 'package' of approximately USD 120,000 to 220,000 (surrogates receive only a fraction of this; see Reuters report by Harney 2013). This is likely to have increased in line with inflation.

Parentless-ness/Statelessness

Birth mothers may or may not be the legal mothers of a surrogate baby. In India a surrogate is not the automatic birth mother, whereas in the United Kingdom she is, regardless of the mode of surrogacy (i.e. genetic input). If the commissioning parent is not the automatic legal parent of an overseas-born surrogate baby in her country of origin such as America and if there is no genetic link to the intended parent(s), then the baby is in effect stateless as no one will be able to register it with a legal parent (Henaghan 2013). For example, some years ago, a case of an American intended mother and a Jamaican father came to the attention of the media (Umanadh 2012). The couple had used his sperm—but the intended father could not fly to India where the surrogacy was arranged. The mother did not provide her own oocytes but donor oocytes were used. Since this child had no genetic link to the mother who was the only one with US citizenship, the courts could not grant the baby her citizenship. After a prolonged and distressing time, US citizenship was granted and the mother could fly the child home. Another German/Indian case took two years to be resolved because Germany did not recognise the birth certificates as there is no legal framework for surrogacy. Eventually this case resulted in the successful adoption of the Indian surrogate baby by the German couple (Henaghan 2013). This parentless and stateless position could have consequences for the child. It is possible that a prolonged battle with the authorities and one or both commissioning parents having to go home before the legal parentage is determined could affect their bonding and attachment to the child and vice versa. Similarly, 'limping' parentage—where only one parent is legally recognised as the parent and the other is not—can impact upon the child's status in gaining access to health care or educational settings (Millbank 2013). Only long-term research on these children and parents will be able to show what the effects of these poorly investigated family building opportunities are.

Attitudes Towards Cross-Border Surrogacy Arrangements

Arvidsson et al. (2015) reported that Swedish people, including Swedes using cross-border surrogacy, see Indian surrogacy as 'morally ambiguous'. Commercialisation and consumerism in surrogate motherhood arrangements provide an inappropriate market in children whether through intercountry adoption or international surrogacy. It is also illegal in many countries, including the United Kingdom (Surrogacy Arrangements Act 1985). As noted earlier, in nearly all cases involving commercial surrogacy (as opposed to altruistic), the surrogate mother is substantially less well-educated and less well-employed than her commissioning intended parent(s) counterparts (van den Akker 2003). Furthermore, in developing countries, the disparity seen between surrogate and intended couples is considerably worse, with those surrogates living in serious poverty. Such living conditions could 'drive' poor uneducated women into surrogacy for purely financial reasons and put them at risk of not giving full informed consent. The consequences of the surrogate motherhood experience could leave them with emotional scarring which may continue for the rest of their lives. This type of poverty-struck surrogate mother is also more likely to be exploited by intermediaries and clinics. There is some evidence that in both intercountry adoption and cross-border surrogacy, poverty is cited as the reason for relinquishing their babies (Cheney 2016). The fact that in both cases intermediaries are involved at substantial costs suggests there is the potential for human trafficking as discussed in relation to intercountry adoption (van Loon 1993). However, as previously noted, banning commercial surrogacy in one country leads to surrogacy service brokers shifting their markets to other countries as was found when India closed its doors to foreign couples (see BioEdge, 7 Nov. 2015).

Comparisons Between Gamete Donation, International Adoption and Cross-Border Surrogacy

One of the few things on which there tends to be strong agreement is that in international adoption the welfare of the child is supported by the provision of accurate record keeping, allowing for birth family searches should they wish to find them. In international surrogacy, there are already problems with accessing accurate birth or genetic information. In Indian surrogate motherhood arrangements for example, neither the gamete donors nor the surrogate mothers appear anywhere on the birth certificates or hospital records (Cheney 2016). According to those present at the International Forum on Intercountry Adoption and Global Surrogacy in The Hague, there was substantial agreement that both the women's and children's rights need to be addressed in Intercountry Adoption and Global Surrogacy (HCCH 1995).

Adoption for adopted children is known to shape their identity (Cheney 2016) and birth families often remain a 'hidden dimension' in intercountry adoption (Cheney 2016, p. 13). Research from third-party gamete donation has also shown collective identity orientations to be lower in adult offspring of gamete donation who are aware of the circumstances of their birth and actively registered on a donor link register in the hope of finding genetic relatives (van den Akker et al. 2015, p. 115). The strength of feeling expressed by donor-conceived adults is, for example, reflected in the quotes in Box below:

> 'Curiosity' doesn't go anywhere near the HUNGER (emphasis original) to find someone I was connected to'.

> 'To see whether we have anything in common' sounds so casual. It is a case of looking for CONNECTION (emphasis original). For me, that was not anything in the zone of curiosity or idle research; it was visceral'.

> 'It is a fundamental quest to find family and get to know them and feel a part of a new family and be accepted by them'.

> 'This is my only chance to find blood relatives.' (van den Akker et al. 2015, p. 115)

Although most surrogate-born children are still relatively young and in adoption the desire of adopted children to trace genetic/birth family has been underestimated, in drawing on the adoption and gamete donation research it is likely that surrogate-born children will experience similar effects. Consequently, whether we consider research on adoption, international adoption, surrogacy (both national and cross-border) or gamete donation, the professional's preferred models are for accurate record keeping and openness (see, e.g. the PROGAR (2016), BASW statements, 20/11/2016). This way the offspring can access accurate genetic and birth (gestational) information. All individuals born and then adopted or conceived via gamete or embryo donation, or commissioned and placed in intended parent families as in surrogacy, have a right to information about their genetic and gestational origins. History has shown that adoption agencies are inundated with requests from adoptees of their birth families (Spiers et al. 2005), suggesting it is critical we draw on these experiences and improve accurate record keeping in relation to gamete, embryo donation and surrogacy because in the long term the children conceived in this way are likely to need that information to help construct their identities (van den Akker et al. 2015).

Cross-border surrogate arrangements may also need more scrutiny than was at first thought to ensure inappropriate baby acquisition does not occur. An unusual but nevertheless real example of surrogate baby commodification comes from Bangkok, where a police raid found nine babies with nine nannies, all commissioned by Mitsutoki Shigeta who fathered these and seven other surrogate babies in the space of a couple of years. The suspected motives were human trafficking and exploitation of children. The police are uncertain who the biological mothers of these babies are, and it is possible the children will never find out. Two further issues are of concern here: firstly, this new family, although apparently financially stable, needs to bring in paid help for the 16 very young babies raising concern for their individual development and opportunities for attachment to a parent figure. Secondly, this type of new family is possible only if the commissioning parent is in the position to pay that many surrogates, thus leaving this kind of excessive procreation unequally into the hands of the sufficiently wealthy only.

Lack of Informed Consent

Lack of informed consent, already covered in other chapters, is more acute in surrogates recruited in developing countries with high poverty and poor educational opportunities for women. The media has provided a good description of the difficulties some surrogates have in consenting with a full understanding of all the facts. Patidta Kusolsang, a Thai surrogate, refused to sign the papers necessary for the baby to acquire a passport to leave the country to live with the commissioning parents. According to the surrogate, she had not understood the original contract, which was written in English and discovered the couple were gay only after the birth of the child. She believed gay parents were not considered 'natural' parents in Thai society and she was upset when the couple refused to talk to her about her visiting the baby. A lack of understanding is therefore detrimental not only to the surrogate mother (as she was prevented from seeing the baby) but also for the commissioning parents (who had to go to court to try to get their baby).

Although one of the fathers was the biological father of the surrogate baby, the egg came from an anonymous donor and not the Thai surrogate. Since Thai law recognises the woman who gives birth to a child as its legal mother, the gay commissioning couple could not leave the country with the baby without the consent of the surrogate. The case is even more complicated because Thai law does not recognise same-sex marriages and also because of the new legislation which outlaws commercial surrogacy that took effect after the surrogate baby was born (Asian Correspondent 2016).

The Legal Status of Surrogacy Across the Globe

With many laws dependent upon ancient dominant religious influences, it is not surprising that many religions have openly voiced their opposition to surrogate motherhood. The Roman Catholic Church, fundamentalist Christians and some Orthodox Jews (Lasker and Borg 1987; Rosner 1983) oppose surrogacy with as much conviction as some feminists (Overall 1987; Purdy 1989) and pro-life groups. Religious influences also dictate acceptability of practices at a societal level. Accepting the potential 'costs' to a surrogate and her family, the commissioning couple and the surrogate child and the society they live in may be too much to 'satisfy the generative needs of infertile couples in this way' (Robertson 1983,

p. 29). Surrogacy therefore is initially perceived as threatening traditional values including a society's religions, undermining what 'the family' means and raises serious culture-specific questions about the meaning of parenthood and motherhood. Conversely, nations with a more liberal and less traditional outlook are generally therefore more in favour of surrogate motherhood than countries steeped in tradition. In many traditionally orientated countries commercial surrogacy is also illegal, but that has not deterred their populations to opt for cross-border surrogacy. In addition to also being considerably cheaper, other reasons for crossing borders are because older and same-sex couples can commission babies via cross-border surrogacy, whereas some intercountry adoption agencies will not place children with non-traditional prospective parents (Sifris 2014). Table 8.1 shows most of the countries where commercial and altruistic surrogacy is allowed and where it is not allowed. As can be seen from the table, some US states, India, the Ukraine (Gamble 2009) and many others allow commercial surrogacy. However, commercial surrogacy is more difficult to monitor, and has resulted in some questionable practices, readily publicised by the press (Scott 2009; Wang 2015).

International Surrogate Motherhood and the Balance of Power

A number of examples of known commercial surrogate arrangements in different countries are described below. All show varying degrees of control of the surrogate, the intended mother or the agents or medical directors. These examples also show the socioeconomic disparities between those commissioning and the commissioned parties. The legal position is generally on the side of the commissioning parents as they are believed to be better placed to secure the welfare of the child because of their higher socioeconomic status.

Australia has been reactive in its regulatory process regarding assisted conception and surrogacy. In May 2016, a decision was made by the parliamentary committee that commercial surrogacy remains illegal and that altruistic surrogacy should be regulated more harmoniously at a national level rather than varied and individually determined by each state

Table 8.1 Listing a number of countries where commercial or altruistic surrogate motherhood arrangements are allowed or explicitly not allowed (up to January 2017)

Country	Allowed	Not allowed
Australia	Altruistic (most states)	Commercial
Belgium	Altruistic	
Bulgaria		Altruistic commercial
Cambodia	Commercial	International arrangements
Canada	Altruistic	
Denmark	Altruistic	
Finland	Altruistic commercial	
France		Altruistic commercial
Germany		Altruistic commercial
Greece	Altruistic	
Hungary		Altruistic commercial
Iceland		Altruistic commercial
India	Commercial	International arrangements
Iran	Altruistic[a]	
Ireland		No laws on surrogacy
Israel	Commercial	Not familial or altruistic
Italy		Altruistic commercial
Japan		Altruistic commercial
The Netherlands	Altruistic	
New Zealand	Altruistic	Commercial
Norway		Altruistic commercial
Pakistan		Altruistic commercial
Portugal	Altruistic	Commercial
Quebec		Altruistic commercial
Russia	Altruistic commercial	
Saudi Arabia		Altruistic commercial
Serbia		Altruistic commercial
South Africa	Altruistic	
Spain		Altruistic commercial
Sweden		Altruistic commercial
Switzerland		No
Thailand	Commercial	International arrangements
Ukraine	Commercial	
United Kingdom	Altruistic	Commercial

(continued)

Table 8.1 (continued)

Country	Allowed	Not allowed
Some US states (Florida/ California)	Commercial	
Vietnam	Altruistic	

[a]In Shia law only

(Cook 2016). The Australian Law Reform Commission is tasked with drafting a new framework, placing the welfare of the child high on its list of priorities. In this way, the committee is asked to consider the merits of including details of all parents on the birth certificate and include a record that surrogacy was used and a means to ensure no national rules are breached in cross-border surrogacy cases. According to Blackburn-Starza (2016), this reform is in line with a recent admission by Baroness Warnock that mistakes were made in the past and that the United Kingdom's position on surrogacy is now outdated. She reports on BBC Radio 4's *Woman's Hour* that children should be told the truth about their origins on birth certificates, and although she does not think the naming of a gamete donor or surrogate is necessary, the noting that gamete donation or surrogacy was used is important so that parents are no longer able to deceive their child(ren). Although this is in progress, the proposition continues to undermine the ample research evidence already discussed showing many individuals brought up by non-birth, non-genetic parents need to know their origins. It is expected that the Australian reform commission will base its recommendations upon the research evidence.

Russia: The Russian position is one that legally (but not in terms of religion, which opposes it) allows genetic and gestational commercial surrogacy for expenses but there is no specification on how much. Married couples and single women can use a surrogate if they are unable to bear a child themselves and surrogate mothers must have had at least one child, be physically and mentally healthy and be between 20 and 35 years old. Interestingly the law here focuses on the financial enforcements, not on the custody of the child—with the surrogate mother retaining all rights to the child regardless of the written contract. Here, therefore, the birth mother is again the legal mother of the child (Rivkin-Fish 2013). Rivkin-Fish (2013) describes the Russian framing of surrogacy as a financial transaction and sets that against the American framing of giving a gift of life. She cites examples of surrogates focusing on fraud detection and how much they can charge for their services as surrogates. At the same time the commissioning mothers simultaneously exert as much control as they can over the surrogate, by controlling what the surrogate can and cannot do. However, ultimately, the power here lies with the surrogate, and the lack of regulation can easily lead to injustice and exploitation. No information on numbers of surrogate births is available. Only Russians able to pay for surrogacy can use this option.

Iran: Iranian views on surrogacy are subject to its religious and cultural laws. In Iran, gestational surrogacy is practised and allowed by married Shiite Muslims who are the majority Muslim groupings in Iran. Here Shiite religious leaders can issue 'fatwas' or religious decrees—verified by parliament that they are consistent with the law—which clinicians can draw on to assist in surrogate pregnancies. Their view is that an embryo or foetus transferred from one womb to another is allowed, although there are issues of kinship and inheritance (Aramesh 2009). The Sunni Muslim population, constituting the majority of Muslims across the world, does not tolerate surrogate motherhood because it involves a man's sperm entering a woman he is not married to. There are no national funds for the treatment of infertility, meaning that by default any infertility treatment and particularly surrogacy which is considerably more expensive (adding the surrogates' fee to the existing IVF fees) is therefore again limited to the wealthier population of Iran. Since Iran still has a very large poor population, there is no shortage of poor women willing to become surrogate mothers (Abbasi-Shavazi et al. 2008). Interestingly, in Iran too there is some controversy over who the 'mother' is. The birth mother is the registered mother of the child and one way around this 'problem' has been to register the birth mother's name as that of the intended mother (Akhoundi and Ardakani 2007). Despite this, the genetic mother is believed to be the real mother through which inheritance is passed on and the surrogate mother is seen more as a wet nurse than a mother, and the surrogate child should not marry her (the surrogate) or her children in the future (Abbasi-Shavazi et al. 2008). Nevertheless, according to the Holy Qur'an, the birth mother should also be the mother. There are groups of Iranian communities where both mothers are considered to be the mother of the surrogate child (Govahi 2007). No information on numbers of surrogate births is available.

India, which was for over a decade a popular destination country for cross-border surrogacy, has seen a recent change in the laws not allowing 'foreign' couples into their surrogacy clinics. Commercial surrogacy is still legal for Indian couples via the draft ART Regulation Bill (MoHFW ICMR 2010). Surrogate mothers are reported to be extremely poor, have no work opportunities and are financially desperate. Because of this they are happy to consider (commercial) surrogacy as employment (Hochschild 2009; Pande 2009; SAMA-Resource Group for Women and Health 2012). The term 'dirty' work has been coined by Pande (2009) referring to an undesirable but necessary occupation, with surrogates 'willing' to take on stigmatising work instead of poorly paid work or no work at all. However this is ethically problematic because no country should have a proportion of its population live in such extreme poverty that they are willing to take on work that has been compared to prostitution. One wrong does not justify another wrong.

In India, full control over the surrogate programmes is with the medical directors of the many clinics carrying out surrogacy as a commercial enterprise. Unlike Russia, here varying levels of control to the commissioning parents or the surrogates are lacking. The control almost always resides fully with the clinicians. This control includes:

- financial
- recruitment
- the conception
- the information and subsequent consent of the surrogates to the conditions of service
- the care of surrogates' activities, food, medications once pregnant
- how often the surrogates' own children were allowed to visit their mother
- the birth process and option of choice
- compliance of the surrogate to care for and breastfeed the baby should the medical director or commissioning parents decide this is necessary
- the commissioning parents accepted into the programmes
- their accommodation, translators and agents
- advice on and access to birth registration processes

In these same clinics in India, surrogates sign over all rights to the baby before they become pregnant, and in many cases they do not receive financial payment until they have handed over the baby—a good incentive to ensuring they keep healthy during the pregnancy and a deterrent to keeping the baby (Saravanan 2013). They may also be expected to care for and breastfeed the baby(ies), putting them at further risk of bonding with the child, with little or no post-natal follow-up of their psychological welfare. This could last for eight weeks, the length of time it can take to obtain the babies' passports. Payment to the surrogate is minimal and delayed or staggered, with no continuing payment in full if there are complications (e.g. miscarriage, abortion or abnormalities) along the way. Only Indian couples able to pay can commission a surrogate baby.

Israel allows surrogacy legally with numerous provisos. Israeli surrogates and commissioning mothers must be of the same religion, commissioning parents must be married, but surrogates should not be married although they must have had children. Commissioning parents are required to use the fathers' sperm, whereas an oocyte donor can be used if the commissioning mother does not have viable oocytes. Payment to surrogates, as in India, is staggered, with full payment due upon relinquishment of a healthy baby. Control over the pregnancy, prenatal testing and whether the surrogate gets to see the baby at birth (or ever) is entirely in the hands of the commissioning couple. Surrogates are strongly encouraged, selected and 'groomed' to detach from the foetus, the pregnancy, the commissioning couple and of course the baby (Teman 2010). Again, only people able to pay can afford to commission a surrogate pregnancy.

Japan is considering a legal framework for regulating assisted conception, although guidelines developed by the Japan Society of Obstetrics and Gynecology—approving sperm but not oocyte donation or surrogacy—do exist (IFFS 2004). The risks associated with the procedures to the surrogate mother and the treatment of surrogates as a reproductive means to an end are prohibited by the Japanese Health Sciences Council Assisted Reproductive Technology Committee. They also state that it is undesirable from the viewpoint of the welfare of the child.

A general population survey of people either informed or not informed about gestational surrogacy (since knowledge of biology is not generally available to all of Japanese society, Minai et al. 2007) highlighted gender and socioeconomic differences in attitudes towards surrogacy. Over 2000 individuals received the information; 1564 did not. The former were spilt into those who self-reported having a 'good' or 'poor' understanding of the information provided. Understanding of assisted conception was strongly related to higher age, education and income. Assisted conception information via the information brochure inhibited decision-making in men, but facilitated this in women. Men who received the information were less approving of both types of surrogacy than men who did not receive the information, as did informed women for gestational surrogacy, but not genetic surrogacy. Currently wealthy Japanese parents are known to travel abroad to commission cross-border surrogate babies.

Canada did not have a framework for surrogate motherhood, although there was evidence that the practice was carried out underground within Canada, and via the United States (Begin 1989). In 1985, the Ontario Law Reform Commission (1985) produced the Human Artificial Reproduction and Related Matters report. They stated that 'surrogacy represents a critical threshold that should not be crossed' (cited in Isaacs and Holt 1987, p. 28). In 1994, a survey of Canadian women's attitudes to surrogacy reported that 41% of Canadian women strongly disapproved and only 3% strongly approved of surrogacy for infertile couples. Disapproval for commercial surrogacy was even stronger (75%, compared to 24% who approved of commercial surrogacy). More than 60% of the women who approved of surrogacy were non-Catholic, showing the strong influence of religion on a population's response to surrogacy. Interestingly, as was found in Japan, women with higher education and income were more likely to approve than those who had lower educational qualifications and a lower income (Krishnan 1994). Those using cross-border surrogacy to circumvent national laws like the Japanese are also only those of sufficient financial affluence to be able to do this.

Nigeria, like most sub-Saharan African countries, does not have a regulatory framework for surrogate motherhood arrangements (Onuoha 2014). Despite this, commercial surrogate motherhood, although not readily accepted by its population, is also practised in Nigeria

(Makinde et al. 2016). Nigerian society, like most other African societies, emphasises the importance of reproductive capacity (Inhorn 2015), and tradition continues to attribute infertility to women. For women, the stigma of infertility and the suffering they may experience can be great. Despite this, acceptance and awareness of assisted conception techniques including surrogacy is limited (Ajayi and Dibose-Osadolor 2011). Furthermore, the increasing awareness of criminal baby factories has blurred boundaries between legal (surrogacy) and illegal (baby factories) methods to obtain a baby for some people and has increased the stigma associated with surrogacy as an option to overcome involuntary childlessness.

Assisted conception in Nigeria is also tainted by the 'black market', surrogacy-like baby factories posing as surrogate motherhood clinics (Makinde et al. 2016). These 'factories' provide a birthing place for young women who are pregnant accidentally or women who have been kidnapped and raped. Once they have given birth, the babies are sold (Makinde et al. 2015) and the evidence links these places to human trafficking rings worldwide (Huntley 2013). According to Huntley, the first cases of 'baby harvesting' were reported in 2006 by the United Nations Organization for Education, Science and Culture in its policy paper 'Human trafficking in Nigeria: Root Causes and Recommendations'. However, the case studies related to 'baby harvesting' examined by the policy paper did not evidence exploitation of persons and would more likely be categorised as the sale of infants or illegal adoptions. The true barbaric horrors of the practice were subsequently described by Onuoha (2014). According to Onuoha, in some of the more extreme cases women are coerced into these places where men function to impregnate them or to kick-start another pregnancy. Apart from obvious human rights offences to these young women and their babies, these women will not be receiving adequate ante- and post-natal care as these factories operate in secret (Makinde 2015). Allegedly, their babies are sold for international or domestic adoptions, rituals, slave labour or sexual exploitation (see Uduma Kalu, cited in Huntley 2013, p. 2.). As with all other countries, only individuals able to pay the huge sums of money involved are able to embark on surrogate motherhood arrangements.

Socioeconomic Disparities

The examples of surrogate motherhood practices in the countries out-lined above have all demonstrated the socioeconomic disparities in opportunities for individuals or couples to access reproductive health care and surrogacy. The majority of the populations resident in these and other countries not described in this chapter would not be able to contemplate using a surrogate motherhood arrangement to build their family because of affordability inequalities. It is only the most driven and wealthier fraction of the populations who are able to pursue this option to build a family. However, even amongst these apparently lucky few, huge emotional and financial stresses and strains taint the practice. Inhorn (2015), an anthropologist and international affairs expert, has published a wealth of outstanding work on what she calls 'repro-travel-lers' (instead of repro tourism, because the journey is not a holiday). In a recent book, she describes the experiences of over 200 people travel-ling to Dubai to seek assisted conception treatment because this is not available, safe, legal or affordable within their home countries. Their experiences are somewhat different from individuals seeking surrogacy because cross-border fertility treatment in her study population of infer-tility treatment was sought to escape the pain, fear and danger they face being childless in their home countries. Nevertheless, both examples—Inhorn's repro-travellers for IVF treatment abroad and commissioning couples seeking cross-border surrogacy—take these steps because their home countries do not cater effectively for their reproductive needs, a failure even more pronounced in many developing countries. All are subjected to substantial emotional and financial hardships in their quests to build a family, showing alternative and affordable means to parent-hood need to be rethought across the globe. A model incorporating prevention as well as responsibility for the outcomes and consequences of reproductive health would prioritise the future of reproductive health treatment from a lifetime and ethically sound perspective (van den Akker 2013).

Commissioning Couples' Needs Versus Exploitation of Commercial Surrogate Mothers and Harm to Surrogate Children

As noted at the start of this chapter, cross-border commercial surrogate mothers are nearly always extremely poor with unemployed husbands (if they have one) and they have difficulty schooling or seeking medical help for their children (Limon 2013). Despite this, a number of studies indicate cross-border commercial surrogate mothers do not report feeling exploited but empowered (Pande 2009; SAMA 2012; Deomampo 2013; Karandikar et al. 2014). However, there are also reports of significant health risks to commercial surrogate mothers particularly in poor resource areas (Bishop and Loff 2013; Bromfield and Smith-Rotabi 2014). In cross-border commercial surrogacy (Goswami et al. 2014), just like in national altruistic surrogacy (van den Akker 2003), sadness following relinquishment of the baby is reported and little is done to support these women with their loss, since no international law secures high standards of counselling care that are met at local level (Engel 2014).

In addition to the need for counselling, (medical) ethical principles— such as to do no harm—are not universally applied in cross-border surrogacy arrangements, and the interests of one party may take precedence over the other even if this could result in serious harm to the other party (the surrogate and the surrogate-born child). For example, Sylvestre-Margolis et al. (2015) reported on the numbers of embryos some gay men transfer in surrogate arrangements. Most gay men were willing to have 2 embryos transferred, 19% 1 embryo and 13% transferred 3+ embryos, because they believed that would give them a greater chance of a successful pregnancy and some reported a desire to have twins (26%). These are also reported as heterosexual couples' reasons for choosing twins or multiples (van den Akker and Purewal 2011; Latar and Razali 2014). Sylvestre-Margolis et al. (2015) also reported a desire to have embryos from each of the male partners in the same-sex relationships (11%), as a reason to have twins or multiples. International monitoring, regulations and laws could ensure the welfare of the (gay) couples, the surrogates and the resultant babies is optimised by educating the parents and surrogates about the dangers of multiple embryo transfers, a recommendation that is a long way from being realised at an international level.

Another serious problem evident in cross-border surrogacy concerns the lack of information shared between the countries' immigration and other government departments such as the justice/legal, and health departments. Cross-border surrogacy arrangements effectively operate within an information vacuum, devoid of national records and potentially of individuals who are 'flagged up' as potentially or actually dangerous. A number of examples already described in this and previous chapters concern the 'baby Gammy' scandal and the Japanese millionaire who wanted to have a large number of children to run his business (Rawlinson 2014). For example, in the baby Gammy case (Schover 2016), the Australian man met his wife (age unknown) on a Chinese Internet site that brokers international marriages. It is known that international marriage brokering sites are associated with human trafficking and domestic abuse of women. The couple then used a broker in Thailand for a gestational surrogate arrangement. There was no information confirming who contributed the oocyte or why the commissioning mother could not carry a pregnancy. The surrogate carried twins—one of them had Down's syndrome—and the surrogate was asked to abort the pregnancy which she refused on religious grounds. The twins were born prematurely, and the couple took the healthy female twin leaving Gammy, the critically ill male twin with the surrogate who was willing to keep him. A lengthy media frenzy reported on this case, resulting in a large sum of money raised to pay for Gammy's medical expenses. The media enquiries also brought to light information about the commissioning father's criminal history—a convicted child molester—his wife knew of his criminal record. Following this very public case of exploitation and commodification, the Thai Government changed the country's laws to ban commercial surrogacy and establish criminal penalties for engaging in or arranging the practice (BBC News Asia 2014). This sudden legal change affected many dozens of babies already created for many people including the nine babies fathered with surrogate mothers by a Japanese millionaire who needed a generation of children to run his business in the future, and many other ongoing pregnancies using Thai surrogate mothers or gestational carriers (BBC News Asia 2014). Similarly,

Chinese couples started bringing Chinese surrogates to Thai clinics and returning with them to China once pregnant, and registering themselves as the parents on the birth certificate.

Equality in Access to Surrogate Mothers

Equality in access to good clinical care to build a family in non-traditional families such as same-sex couples is also patchy. In 2015, Smotrich et al. retrospectively reviewed IVF laboratory databases and patients' charts over a 10-year period. A total of 529 consecutive fresh cycles of gestational surrogate arrangements using egg donors were performed for gay couples, and 80% achieved a live birth after one cycle. The gay couples had travelled from 54 countries and 47 US states. The majority had learned about gestational surrogacy via the internet and the media and nearly all (96%) stated that official reports such as Clinic Success Rates were not available to them, suggesting although gestational surrogacy using egg donation is clearly successful for gay couples they do not have sufficient access to official statistical data specifically for LGBT patients in order to make fully informed decisions about the best treatment facility for their needs.

Cross-border surrogacy practices are not always in the best interests of the children or the surrogate mothers as fears of abuse, trafficking and exploitation are a reality, albeit not common. For many commissioning parents, cross-border fertility services continue to allow them the opportunity to circumvent international laws (Engel 2014). That was also demonstrated in a study by Rodino et al. (2014). They analysed anonymous questionnaires from 137 Australian and New Zealand residents who were recruited from Internet sites on cross-border reproductive services. Surprisingly, local laws prohibiting various types of ART, such as commercial surrogacy or sex selection for family balancing or statutes restricting access to ART to heterosexual couples, did not deter people from pursuing this cross-border option. Furthermore, most respondents believed cross-border reproductive services were medically sound (91.2%), safe (89.4%) and financially reasonable (85.7%), but of the relatively small number who actually pursued treatment, only approximately half reported that their emotional needs were met. Thorn et al. (2012) have advocated an internationally agreed model for ethically based minimum standards of care service provision to ensure future surrogate motherhood provision is transparent and accountable.

Summary

In summary, cross-border surrogacy brings with it many additional problems amplifying those evident in national arrangements, in part because of the availability of cheaper surrogates and IVF clinics, the lack of screening of commissioning parents and a relaxed attitude to give people what they want (such as sex selection or multiples or twins). The welfare of the surrogate mother and that of the surrogate baby may be compromised. In the long term the surrogate mother and the baby are at risk of exploitation and commodification, although commissioning parents too may be at the mercy of surrogates' demand or refusal to hand over the baby. Finally, the lack of effective legislation in the countries used for cross-border arrangements poses huge nationality and citizen problems, potentially resulting in delays in forming secure relationships between the babies and the commissioning parents. There is a total lack of information on the long-term welfare of the triads involved in international surrogacy. The next chapter considers the main moral, ethical and human rights issues posed by surrogate motherhood arrangements.

References

Abbasi-Shavazi, M. J., Razeghi-Nazrabad, H. B., & Toloo, G. (2008). The Iranian ART revolution: Infertility, assisted reproductive technology and third party donation in the Islamic Republic of Iran. *Journal of Middle East Women's Studies, 4*(2), 1–28.

Ahmad, T. (2015). India: Draft legislation regulating assisted reproductive technology published. 2 Nov 2015. http://www.loc.gov/law/foreign-news/article/india-draft-legislation-regulating-assisted-reproductive-technology-published/. Accessed 21 Mar 2017.

Ainsworth, M. (1990). Some considerations regarding theory and assessment relevant to attachments beyond infancy. In M. T. Greenberg, D. Cicchetti, & E. M. Cummings (Eds.), *Attachment in the preschool years: Theory, research and intervention* (pp. 463–488). Chicago: University of Chicago Press.

Ajayi, R. A., & Dibosa-Osadolor, O. J. (2011). Stakeholders views on ethical issues in the practice of in-vitro-fertilisation and embryo transfer in Nigeria. *African Journal of Reproductive Health, 15*, 73–80.

Akhoundi, M. M., & Ardakani, Z. B. (2007). Surrogacy, definition, types and its necessity in treatment of infertility. (In Farsy) In *Medical, legal, Islamic jurisprudential, ethical, philosophical, sociological and psychological aspects of surrogacy* (pp. 3–14). Theran: Samt.

Ali, L., & Kelly, A. (2008). The curious lives of surrogates. *Newsweek*. http://www.newsweek.com/curious-lives-surrogates-84469. Accessed 25 June 2014.

Aramesh, K. (2009). Iran's experience with surrogate motherhood: An Islamic view and ethical concerns. *Journal of Medical Ethics, 35*, 320–322.

Arvidsson, A., Johnsdotter, S., & Essen, B. (2015). Views of Swedish commissioning parents relating to the exploitation discourse using transnational surrogacy. *PloS One, 10*(15), e126518. doi:10.1371/journal.none.0126518.

Asian Correspondent Staff. (2016, March 24). https://asiancorrespondent.com/2016/03/gay-couple-begin-legal-battle-in-thailand-against-surrogate-over-custody-of-baby/. Accessed 16 April 2016.

BBC Channel 4. (2013). Panorama 'house of surrogates'. http://www.bbc.co.uk/programmes/b03c591s

BBC News Asia. Thai surrogate baby Gammy: Australian parents contacted. http://www.bbc.com/news/world-asia-28686114. Accessed 8 Aug 2014.

Begin, P. (1989). *New reproductive technologies*. Ottawa: Research Branch, Library of Parliament.

BioEdge. (2015, November 7). Surrogacy business shifts to Cambodia. http://bioedge.org/bioethics/surrogacy-business-shifts-to-cambodia/11638

Bishop, L., & Loff, B. (2013). The rights of the gestational mother and child in surrogacy: A bill to regulate surrogacy in India. *Australian Journal of Adoption, 7*(3), 1–8.

Blackburn-Starza, A. (2016, May 9). Australian parliamentary committee calls for national surrogacy regulation. *BioNews*. http://www.bionews.org.uk/page_646844.asp. Accessed 25 Oct 2016.

Blyth, E. (1994). "I wanted to be interesting, I wanted to be able to say 'I've done something with my life'": Interviews with surrogate mothers in Britain. *Journal of Reproductive and Infant Psychology, 12*, 189–198.

Blyth, E., Crawshaw, M., & van den Akker, O. (2014, February 17). What are the best interests of the child in international surrogacy? *Bionews*, p. 742.

Bowlby, J. (1969/1982). *Attachment and loss: Vol I Attachment*. New York: Basic Books.

Bromfield, N., & Capous-Desyllas, M. (2012). Underlying motives, moral agendas and unlikely partnerships: The formulation of the U.S. trafficking victims protection act through the data and voices of key policy players. *Advances in Social Work, 13*(2), 243–261.

Bromfield, N., & Smith-Rotabi, K. (2014). Global surrogacy, exploitation, human rights and international private law: A pragmatic stance and policy recommendations. *Global Social Welfare, 1*(3), 123–135.

Carney, S. (2010). Cash on delivery: Gestational dormitories. Routine C-sections. Quintuple embryo implants. Brave new world? Nope surrogacy tourism. *Mother Jones*, pp. 69–73. http://www.motherjones.com/politics/2010/02/surrogacy-tourism-india-nayna-patel. Accessed 21 Mar 2017.

Cheney, K. (2016). Preventing exploitation, promoting equity: Findings from the international forum on intercountry adoption and global surrogacy 2014. *Adoption & Fostering, 40*(1), 6–19.

Ciccarelli, J., & Beckman, L. (2005). Navigating rough waters: An overview of psychological aspects of surrogacy. *Journal of Social Issues, 61*(1), 21–43.

Cook, M. (2015, November 7). Surrogacy business shifts to Cambodia. *BioEdge*.

Cook, M. (2016, May 7). Australian government report backs ban on commercial surrogacy. *BioEdge*. http://www.bioedge.org/bioethics/australian-govt-report-backs-ban-on-domestic-commercial-surrogacy/11864. Accessed 12 May 2016.

Corea, G. (1985). *The mother machine: Reproductive technologies from artificial insemination to artificial wombs*. New York: Harper & Row Publishers.

Crawford, T., Cohen, P., Chen, H., Anglin, D., & Ehrensaft, M. (2009). Early maternal separation and the trajectory of borderline personality disorder symptoms. *Development and Psychopathology, 21*, 1013–1030.

Crawshaw, M., Blyth, E., & van den Akker, O. (2013a). The changing profile of surrogacy in the UK: Implications for national and international policy and practice. *Journal of Social Welfare and Family Law, 34*(3), 1–11.

Crawshaw, M., Purewal, S., & van den Akker, O. (2013b). Working at the margins: The views and experiences of court social workers on parental orders' work in surrogacy arrangements. *British Journal of Social Work, 43*, 1225–1243.

Cuthbert, D., & Fronek, P. (2013). Perfecting adoption? Reflections on the rise of commercial offshore surrogacy and family formation in Australia.

In *Families, policy and the law*. Melbourne: Australian Institute of Family Studies/Australian Government.

Deomampo, D. (2013). Transnational surrogacy in India: Interrogating power and women's agency. *Frontiers: Journal of Women's Studies, 34*(3), 167–188.

Engel, M. (2014). Cross-border surrogacy: Time for a convention? In K. Boele-Woelki, N. Dethloff, & W. Gephert (Eds.), *Family law and culture in Europe: Developments, challenges, and opportunities, European family law* (Vol. 35, pp. 199–216). Cambridge, UK: Intersentia Ltd.

Fixmer-Oraiz, N. (2013). Speaking of solidarity: Transnational gestational surrogacy and the rhetorics of reproductive (in)justice. *Frontiers, 43*, 126–163.

Gamble, N. (2009). Crossing the line: The legal and ethical problems of foreign surrogacy. *Reproductive Biomedicine Online, 19*(2), 151–152.

Goswami, L., Rotabi, K., & Bromfield, N. (2014). *Force, fraud, coercion and exploitation: Abuses in intercountry adoption contrasted against what Indian women acting as surrogate mothers have to say about their experiences.* Paper presented at the Forum on Intercountry Adoption and Global Surrogacy. The Hague, 11–13 August.

Govahi, Z. (2007). Assessment of jurisprudential aspects of surrogacy. In *Medical, legal, Islamic jurisprudential, ethical, philosophical, sociological and psychological aspects of surrogacy* (pp. 122–138). Samt: Theran.

Gugucheva, M. (2010). *Surrogacy in America.* Council for Responsible Genetics. http://www.councilforresponsiblegenetics.org/pagedocuments/kaevej0a1m.pdf

Harney, A. (2013). Why wealthy Chinese are seeking U.S. surrogates. Reuters. http://www.reuters.com/article/2013/09/22/us-china-surrogates-idUSBRE98L0JD20130922. Accessed 21 Mar 2017.

HCCH. (1995). Hague Conference on Private International Law 33: The Hague Convention of 29 May 1993 on protection of children and co-operation in respect of inter country adoption (Hague Adoption Convention). http://hcch.net/index_en.php?act=conventions.text&cid=69. Accessed 22 May 2016.

Henaghan, M. (2013). International surrogacy trends: How family law is coping. *Australian Journal of Adoption, 7*(3), 1–24.

Hochschild, A. (2009). Childbirth at the global crossroads. *The American prospect.* http://prospect.org/article/childbirth-global-crossroads-0. Accessed 23 May 2016.

Huntley, S. (2013). The phenomenon of 'baby factories' in Nigeria as a new trend of human trafficking. *International Crimes Database.* Brief 3. http://www.internationalcrimesdatabase.org/upload/documents/20131030T045906-ICD%20Brief%203%20-%20Huntley.pdf. Accessed 12 May 2016.

IFFS. (2004). International Federation of Fertility Societies International Conference. IFFS surveillance 04. *Fertility and Sterility, 81*(5, 4), 1S–54S.

Inhorn, M. (2015). *Cosmopolitan conceptions: IVF sojourns in Dubai*. Durham: Duke University Press.

International Social Service. (2014, May). The adoption of older children: A project that measures up to the children's needs? (First part). *ISS Monthly Review, 181*.

Isaacs, S. L., & Holt, R. J. (1987). Redefining procreation: Facing the issues. *Population Bulletin, 42*(3), 1–37.

Karandikar, S., Gezinski, L., Carter, J., & Kalonga, M. (2014). Economic necessity or noble cause? A qualitative study exploring motivations for gestational surrogacy in Gujarat, India. *Affilia, 29*(2), 224–236.

Kobak, R., & Marsden, S. D. (2008). The emotional dynamics of disruptions in attachment relationships: Implications for theory, research and clinical intervention. In J. Cassidy & P. R. Shaver (Eds.), *Handbook of attachment* (Vol. 2, pp. 23–47). New York: Guilford Press.

Kohli, N. (2011). Mom's market. *Hindustan Times*. Delhi: Hindustan Times. http://www.hindustantimes.com/india/moms-market/story-4V8X4IoEq LMRO82K3x50XI.html. Accessed 21 Mar 2017.

Krishnan, V. (1994). Attitudes toward surrogate motherhood in Canada. *Health Care for Women International, 15*, 333–357.

Lasker, J., & Borg, S. (1987). *In search of parenthood*. Boston: Beacon Press.

Latar, I., & Razali, N. (2014). The desire for multiple pregnancy among patients with infertility and their partners. *International Journal of Reproductive Medicine*. doi:10.1155/2014/301452.

Limon, C. (2013). Surrogacy and parenthood: An overview of the research on the relationship between surrogacy and adoption. *Australian Journal of Adoption, 7*(3), 1–6.

Makinde, O. A. (2015) Infant trafficking and baby factories: A new fate of child abuse in Nigeria. *Child Abuse Review*. Doi:10.1002/car.2420. Accessed 12 May 2016.

Makinde, O. A., Olaleye, O., Makinde, O. O., Huntley, S., & Brown, B. (2015). Baby factories in Nigeria: Starting the discussion toward a national prevention policy. *Trauma, Violence & Abuse*. Doi:10.1177/1524838015591588. Accessed 12 May 2016.

Makinde, O. A., Makinde, O. O., Olaleye, O., & Odimegwu, C. O. (2016). Baby factories taint surrogacy in Nigeria. *Reproductive Biomedicine Online, 32*(1), 608.

Millbank, J. (2013). Resolving the dilemma of legal parentage for Australians engaged in international surrogacy. *Australian Journal of Family Law, 27,* 136.

Minai, J., Suzuki, K., Takeda, Y., Hoshi, K., & Yamagata, Z. (2007). There are gender differences in attitudes toward surrogacy when information on this technique is provided. *European Journal of Obstetrics and Gynecology and Reproductive Biology, 132,* 193–199.

MoHFW ICMR. (2010). *The assisted reproductive technology (regulation) bill (draft).* New Delhi: Ministry of Health and Family Welfare/Indian Council of Medical Research.

Ontario Law Reform Commission (OLRC). (1985). *Report on human artificial reproduction and related matters* (Vol. I and II). Toronto: Ministry of the Attorney General.

Onuoha, F. C. (2014). The evolving menace of baby factories and trafficking in Nigeria. *African Security Review, 23,* 405–411.

Overall, C. (Ed.). (1987). *The future of human reproduction.* Toronto: The Women's Press.

Pande, A. (2009). Not 'an angel' not 'a whore': Surrogates as 'dirty workers' in India. *Indian Journal of Gender Studies, 16*(2), 141–173.

Pande, A. (2010). At least I am not sleeping with anyone: Resisting the stigma of commercial surrogacy in India. *Feminist Studies, 36*(2), 292.

Progar. (2016, November 11). Progar BASW surrogacy position paper. https://www.basw.co.uk/progar/

Purdy, L. M. (1989). Surrogate mothering: Exploitation or empowerment? *Bioethics, 3,* 18–34.

Rawlinson, K. (2014). Interpol investigates 'baby factory' as man fathers 16 surrogate children. *The Guardian.* https://www.theguardian.com/life-andstyle/2014/aug/23/interpol-japanese-baby-factory-man-fathered-16-children. Accessed 25 Oct 2016.

Rimm, J. (2009). Booming baby business: Regulating commercial surrogacy in India. *University of Pennsylvania Journal of International Law, 30,* 1429–1462.

Rivkin-Fish, M. (2013). Conceptualizing feminist strategies to Russian reproductive politics: Abortion, surrogate motherhood, and family support after socialism. *SIGNS, Journal of Women in Culture and Society, 38*(3), 569–594.

Robertson, F. (1983). Surrogate mothers: Not so novel after all. *The Hastings Center Report, 13*(5), 28–34.

Rodino, I. S., Goedeke, S., & Nowoweiski, S. (2014). Motivations and experiences of patients seeking cross border reproductive care: The Australian and New Zealand context. *Fertility & Sterility, 102*(5), 1422–1431.

Rosner, F. (1983). In vitro fertilization and surrogate motherhood: The Jewish view. *Journal of Religion and Health, 22*(2), 139–159.

SAMA-Resource Group for Women and Health. (2012). Birthing a market: A study on commercial surrogacy. http://www.samawomenshealth.org/downloads/birthing%20A%Market.pdf. Accessed 23 May 2015.

Saravanan, S. (2013). An ethnomethodological approach to examine exploitation in the context of capacity, trust and experience of commercial surrogacy in India. *Philosophy, Ethics and Humanities in Medicine, 8*(10), 1–12.

Schover, L. (2016). Cross-border surrogacy: The case of Baby Gammy highlights the need for global agreement on protections for all parties. *Fertility & Sterility, 102*(5), 1258–1259.

Scott, E. S. (2009). Surrogacy and the politics of commodification. *Law and Contemporary Problems, 72*, 109–146.

Sherman, R., Misca, G., Rotabi, K., & Selman, P. (2016). Global commercial surrogacy and international adoption: Parallels and differences. *Adoption & Fostering, 40*(1), 20–35.

Sifris, A. (2014). Gay and lesbian parenting: The legislative response. In *Families, policy and the law*. Melbourne: Australian Institute of Family Studies/ Australian Government.

Smotrich, D., Botes, A., Wang, X., Gaona, M., & Batzofin, D. (2015). Gay surrogacy-the quandry of accessing verifiable facts. *Fertility & Sterility, 104*(3), e33.

Speirs, C., Duder, S., Sullivan, R., Kirstein, S., Propst, M., & Meade, D. (2005). Mediated reunions in adoption: Findings from an evaluation study. *Child Welfare, 84*(6), 843–866.

Stark, B. (2012). Transnational surrogacy and international human rights. *ILSA Journal of International & Comparative Law, 18*(2), 1–16.

Surrogacy Arrangements Act. (1985). Chapter 49. http://www.legislation.gov.uk/ukpga/1985/49. Accessed 18 June 2015.

Surrogacy matters: Inquiry into the regulatory and legislative aspects of international and domestic surrogacy arrangements. (2016, May 4). Parliament of Australia. http://www.aph.gov.au/Parliamentary_Business/Committees/House/Social_Policy_and_Legal_Affairs/Inquiry_into_surrogacy/Report. Accessed 12 May 2016.

Sylvestre-Margolis, G., Vallejo, A., & Rauch, E. (2015). Gestational surrogacy/egg donor IVF: Behavior of gay men intended parents with respect to numbers of embryos transferred. *Fertility and Sterility, 104*(3), e57. doi:10.1016/j.fertnstert.2015.07.173.

Symons, X. (2015, June 20). Nepalese surrogacy unearthed. *BioEdge.*

Teman, E. (2008). The social construction of surrogacy research: An anthropological critique of the psychosocial scholarship on surrogate motherhood. *Social Science & Medicine, 67*(7), 1104–1112.

Teman, E. (2010). *Birthing a mother: The surrogate body and the pregnant self.* Berkeley: University of California Press. isbn:978–0–520-25964-5.

Thorn, P., Wischmann, T., & Blyth, E. (2012). Cross-border reproductive services – Suggestions for ethically based minimum standards of care in Europe. *Journal of Psychosomatic Obstetrics and Gynecology, 33*(1), 1–16.

Umanadh, J. (2012). MEA may allow American woman to take home surrogate child. *Deccan Herald.* http://www.deccanherald.com/content/222594/mea-may-allow-american-woman.html. Accessed 21 Mar 2017.

van den Akker, O. (2001). Adoption in the age of reproductive technology. *Journal of Reproductive and Infant Psychology, 19*(2), 147–159.

van den Akker, O. (2003). Genetic and gestational surrogate mothers' experience of surrogacy. *Journal of Reproductive and Infant Psychology, 21*(2), 145–161.

van den Akker, O. (2007). Psychosocial aspects of surrogate motherhood. *Human Reproduction Update, 13*(1), 53–62.

van den Akker, O. B. A. (2013). For your eyes only: Bio-behavioural and psychosocial research objectives. *Human Fertility, 16*(1), 89–93.

van den Akker, O., & Purewal, S. (2011). Elective single-embryo transfer: Persuasive communication strategies can affect choice in a young British population. *Reproductive Biomedicine Online, 23*(7), 838–850.

van den Akker, O. B. A., Crawshaw, M. C., Blyth, E. D., & Frith, L. J. (2015). Expectations and experiences of gamete donors and donor-conceived adults searching for genetic relatives using DNA linking through a voluntary register. *Human Reproduction, 30*(1), 111–121.

van Loon, H. (1993). Report on intercountry adoption, preliminary Document No 1. of April 1990. In Hague conference on private international law, proceedings of the seventeenth session (1993), Tome II, Adoption – co-operation, pp. 11–119.

Wang, P. (2015). *Thailand bans commercial surrogacy for foreigners, singles.* Bangkok: The Associated Press. www.businessinsider.com/ap-thailand-bans-commercial-surrogacy-for-foreigners-singles-2015-8?IR=T. Accessed 21 Mar 2017.

9

Ethical, Moral and Human Rights Considerations in Surrogate Motherhood

In this chapter, ethical and moral principles and issues of human rights are explored in the context of surrogate motherhood arrangements. It draws upon different conceptualisations of surrogacy as a social, legal or medical contract. The ethical problems associated with contracting pregnancies are discussed as they apply to accountability, moral obligations, social and psychological aspects of parenting and in relation to maintaining optimum human rights principles for the commissioned children, surrogate mothers and commissioning parents. The commodification and exploitation known to be practised in the twenty-first century is likely to come to haunt the offspring in later years. The complications arising from national guidelines which clarify local, but obfuscate international, surrogate motherhood arrangements are also explored.

The Universal Declaration of Human Rights 1948, Article 1 states: All human beings are born free and equal in dignity and rights. They are endowed with reason and conscience and should act towards one another in a spirit of brotherhood (UN General Assembly 1948). The Universal Declaration on Bioethics and Human Rights (UNESCO 2006) incorporates an acknowledgement of technological advances in medical science and states they should be ethically sound; 'giving due respect to the dignity of the human person and universal respect for, and observance

© The Author(s) 2017
Olga B.A. van den Akker, *Surrogate Motherhood Families*,
DOI 10.1007/978-3-319-60453-4_9

of, human rights and fundamental freedoms'. A Kantian moral perspective as a foundation for a human rights and dignity perspective (thereby avoiding commodification of the body in cross-border surrogacy) may be useful in conceptualising the protection of those with least power in these arrangements (Galloway 2015). Communities, justice, power and the social interactions between people require cooperation from the society to which it is applied (Giddens 1984). Surrogacy takes place in many diverse communities across the world (see Chap. 8) and has increased exponentially in recent years (Crawshaw et al. 2012), particularly in countries jostling for economic development and this has led to concerns of social justice, exploitation and human rights breaches (Bromfield and Rotabi 2014).

In countries which allow surrogacy, surrogate motherhood has been conveniently conceptualised as a purely social contract (Saravanan 2013) even when medical intervention (as in gestational surrogacy) is implemented. This is problematic in ethical terms for a number of reasons. Firstly, the fact that medical technology is involved and is not accountable in any way for any aspect of the arrangement seems contrary to medical ethical principles. At the same time it is equally difficult to see how they could be held to account ethically because the arrangements are in some countries like the United Kingdom put together via social contracts. Social contracts only serve to laying out the terms of the agreement in principle without obligation to either party to fulfil the terms of the contract. It creates a moral obligation, not a formal agreement. Lawyers may be consulted to draw up contracts but even these are legally not binding because the law states the birth mother is the legal mother. In other countries such as India, the clinics are involved in the full brokering process and they adhere to their own national laws determining legal parenthood. Currently in some nations national guidelines exist which clarify local, but obfuscate international, surrogate motherhood arrangements. There are no universally applied guidelines or regulations which disentangle cross-border surrogacy, although the arrangements between the parties are universally seen as social contracts (see also Chaps. 8 and 10).

Baier's (1986) framework of social relationships as involving 'trust' (where justice and exploitation can thrive in equal measure) and 'power' (conceptualised within the theory of structuration and Foucault's (1980) biopower) are relevant in debates about the ethics of surrogate motherhood. Trust and (bio)power within surrogate motherhood arrangements seriously put into question the actual, not assumed, human rights exercised in surrogate motherhood arrangements. For example, power, however defined even within social contracts, is nearly always under patriarchal control and does little to strengthen the liberty of the women within surrogate motherhood arrangements (Pateman 1988). According to Held (1993) the social interactions in surrogacy within social contract theory do not represent the women or the children sufficiently and neglect the moral relationships between the people involved in these social contracts. It also fails to capture the interdependent relations (Baier 1994) and trust puts the trustee in a vulnerable position (Baier 1986).

Despite the existence of ethical and moral problems of power and trust in surrogate motherhood arrangements and the context of a flimsy social contract, most surrogate arrangements are carried out without legal involvement (Charo 1988). At a sociopsychological level, however, there are moral and ethical risks which are not adequately dealt with within any system: the legal, medical or health care systems (see Part I, the introductory Chap. 1, and Part II, Chaps. 4 and 5 on surrogate mothers and commissioning parents respectively). The fact that the overwhelming majority of commissioning parents are middle class, well educated, well off and aged in their 30s and 40s and the overwhelming majority of surrogate mothers are working/lower class, not well educated (or illiterate), not well off financially or on the poverty line and aged in their 20s and 30s shows the social relationship is unequally distributed on sociodemographic characteristics. Trust is necessary but power will shift between these parties, leading to different vulnerabilities which, although now well recognised, are not well catered for in the societies which participate in these arrangements. According to Foucault (1980, p. 56.), 'power, although not entirely without its benefits to the person without the power, can retreat here, re-organise its forces, invest itself elsewhere and so the battle continues.'

Legal issues in surrogate motherhood arrangements mainly arise when the baby is born and needs to be registered (Crawshaw et al. 2012) and when the surrogate mother fails to relinquish the baby to the individual or couple commissioning it, or where the commissioning parent(s) refuse to take the baby. The earliest and very well-known case of 'Baby M' (Rothenberg 1988) involved the non-relinquishment of a gestational surrogate arrangement where the judge allowed custody to the genetic commissioning parents in preference over the birth (legal) mother—the surrogate. A similar case involved a ruling by a judge (Oxman 1993), both cases setting a precedent for genetic over birth motherhood, and ruling against local laws. Between countries, the fact that laws differ in different American states (Andrews and Elster 2000) and across the world (Leeton 1991) is no deterrent to bypassing local laws, since commissioning parents simply move around—settling where legislation favours their desired outcome (Johnson 1999). This practice is now prolific across continents. However, it is a practice for those who can afford it. Inequality in surrogate motherhood is therefore rife even amongst those commissioning babies and will be discussed later in this and other relevant chapters.

In any form of commercial surrogate motherhood arrangements where money exchanges hands in return for the creation/production and delivery of a baby, concerns about the ethics and morality of the practice are raised. In surrogate motherhood it is not uncommon for additional parties to be involved, although they may not be directly recruited to or benefit from the surrogate arrangement such as egg donors and/or sperm donors (Rimm 2009; Hague Conference on International Private Law 2012). Agents, agencies and clinics, on the other hand, do benefit from their involvement, particularly in commercial and cross-border surrogate motherhood arrangements and their involvement needs ethical and moral scrutiny. It is important to accept the practice may have a place in modern society with fewer babies available for adoption, the infertility treatment possibilities taking on routine proportions, gamete and embryo donations prolific and the changing nature of marriages and families worldwide. It is imperative to explore the costs and benefits of surrogate motherhood with a view to approaching the subject from a humanistic and dignified perspective. Needless to say, some form of regulation from international law is likely to be necessary via, for example, a Hague Convention on inter-country surrogacy agreements which could provide a working framework for a smoother, ethically acceptable and legally appropriate system of cross-border surrogate arrangements (Ramskold and Posner 2013).

Attitudes to the morality and ethical acceptability of surrogate motherhood are also influenced by national governmental policies and laws. UK law (Surrogacy Arrangements Act 1985) states that surrogacy should be consensual and should not involve payment of more than reasonable expenses. The commercial brokering of surrogacy arrangements is a criminal offence and so is advertising for or by surrogates. Italy, Germany, Singapore, Japan, some Australian states and Turkey's legislation strictly forbid third-party reproduction (Baykal et al. 2008). Population attitudes towards surrogacy of prohibiting countries are likely to be negative as it is criminalised, and this is likely to affect their intention to commission a surrogate or to become a surrogate. Other countries, such as some states in the United States, India, Thailand and the Ukraine, allowed fully enforceable commercial surrogacy (mostly international) arrangements (Gamble 2009). As a result, individuals visiting or living in these permissive (of commercial surrogacy) countries are likely to see surrogacy as an opportunity to parenthood; others as an opportunity to make money, although even here the population attitudes may not always condone the practice potentially disadvantaging those embroiled within it (see also Part I and Chap. 2). Furthermore, some countries which started off permissive such as India, Thailand and Cambodia (which operated in the absence of any form of regulation regarding surrogacy) have now closed their doors to same-sex single and foreign international commissioning parents.

Erlen and Holzman (1990) described the existence of two main approaches to legislation on surrogacy: (1) a prohibitory approach (i.e. a ban on the arrangement) and (2) a regulatory approach or codification of surrogate contracts. Even where a regulatory approach predominates, courts may decide the legality of these laws and make decisions which go against the law, such as, for example, awarding parental custody or parental rights to the commissioning couple if that is considered to be in the best interests of the child even if the law states a birth mother is the legal mother of the child. Since laws are changing and appeals to decisions are becoming more common, increasing changes can be expected within a country's legal frameworks. In order for laws to reflect the needs of its populations, it is critical that those involved in reforms are fully aware of the importance of the biological/medical aspects of surrogacy, as well as the societal and emotional impact on all parties within these societies. They will need to bear in mind not only traditions, cultures and religions dominant in different countries and regions, but also shifts in attitudes and in non-traditional family structures. Finally, and perhaps most importantly, is the ethical and moral issue of legalisation of commodifying children, trading in parts (gametes) or embryos and weighing up of financially beneficial 'offers' of surrogate services and donors' needs to be evaluated. The legal role of professionals offering and liaising between surrogate services, including the duties of lawyers, doctors (Healey 1984), social workers, nurses, midwives, health visitors, counsellors, psychologists and psychiatrists, needs to be understood. Record keeping and identification, as was abundantly obvious in adoption, is now also critically important in surrogacy and gamete/embryo donation, and should ideally form part of any legal statute (Andrews 1987; Elias and Annas 1986; Crawshaw et al. 2012). There has been a tectonic shift in which infertility and ethics are interpreted and the possibilities in new innovative technologies and manipulations in the next decade are likely to be even more challenging requiring socially responsible implementation (Brezina and Zhao 2012).

The Ethical Position

Ethical concerns can be manifold, and medical ethics which are relevant here is based upon four basic principles as shown in Fig. 9.1 (Beauchamp and Childress 1994):

In surrogate motherhood arrangements, one of the most ethically controversial aspects concerns the commercialisation of the process. Ethical principles would focus on the surrogate mother's autonomy, and weigh up the extent to which beneficence and non-maleficence are applicable. The debates, not yet resolved, centre on feminist principles of surrogate mothers being used as incubators leasing a womb (Hadfield 1995; Radin 1987), versus a woman's own opinion on how her dignity is or is not affected by this choice (Erin and Harris 1991; Aramesh 2009). Having

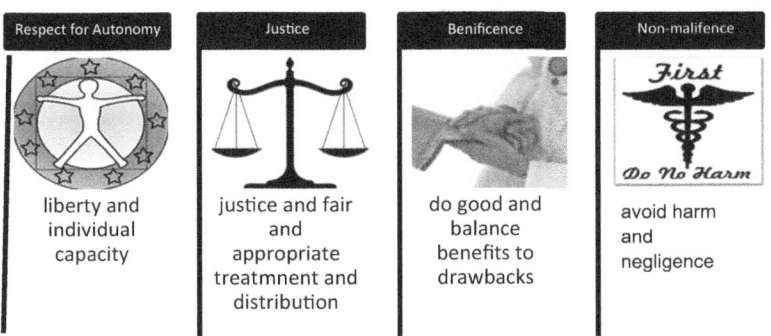

Fig. 9.1 Showing the four basic ethical principles

autonomy and individual freedom is critical (Damelio and Sorensen 2008; van Zyl and Niekerk 2000; Hope et al. 2003). However, surrogate and particularly cross-border surrogate motherhood is based upon socioeconomic, educational and 'status in society' differences, placing surrogate mothers in the vast majority of cases in a vulnerable or disadvantaged position to the commissioning individuals. Consenting to surrogacy in these cases is therefore described as acts of desperation rather than free will and autonomy (Donchin 2010; Damelio and Sorensen 2008; Humbyrd 2009). Clearly regulating this kind of surrogacy is controversial; on the one hand regulation may do away with an individual's liberty and autonomy; on the other hand unregulated surrogacy endangers the basic human rights. The threat to human rights concerns injustice, exploitation, coercion, poverty, dominance and pressure benefitting only the commissioning parents or brokering agents at the expense of the surrogate—and the eventual child brought about through such humanly unacceptable means.

Some assurance that a surrogate mother behaves in an ethically advantageous manner for the unborn child she is influencing is also necessary. There are concerns that some surrogates do not take the care they need of themselves during a surrogate pregnancy (Banjerjee and Basu 2009), thereby potentially compromising the health of the unborn child who had no part to play in the decisions of its own mode of conception. Moral standpoints feed into ethical debates too. Of the major religions

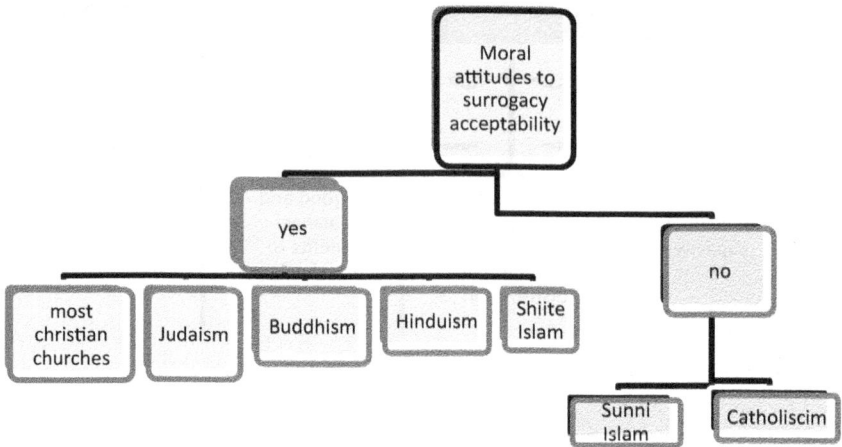

Fig. 9.2 Showing religions with known moral objections/permissions of surrogacy

across the world, some do and some do not consider surrogacy acceptable (see, e.g. Fig. 9.2), and their views will influence its population attitudes towards the surrogacy arrangements.

Ethical and Moral Decision-Making About Gametes and Embryos

Gametes and embryos are obtained from the women, men and couples at some stage for treatment or for donation to others. These gametes and embryos are stored (cryopreserved) for later staggered use. Opinions as to the status of human sperm and oocytes, as well as embryos, have been debated by churches and governments. Their status also matters to the individuals who need to make a decision as to what happens to them once they become obsolete (treatment was successful or continued to fail) to those they were originally intended for. The status attributed to gametes and embryos is diverse (Purewal and van den Akker 2007). However, other factors are also important, including people's experiences of the treatment route, other life circumstances, parenthood status, information and knowledge, as well as demographic and psychosocial factors (de Lacey 2005; Lyerly et al. 2006; Mohler-Kuo et al. 2009; Nachtigall et al. 2009;

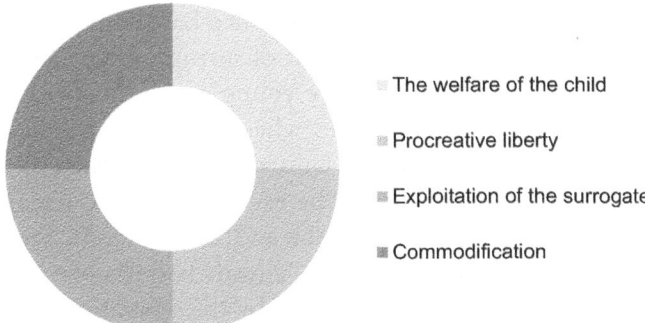

The welfare of the child

Procreative liberty

Exploitation of the surrogate

Commodification

Fig. 9.3 The main ethical problems identified in surrogate motherhood

Provoost et al. 2009, 2012; Jin et al. 2013; Samorinha et al. 2014). A recent study (Bruno et al. 2016) reported parents struggled emotionally with discontinuing cryopreservation of their embryos. They also reported couples were more likely to stop cryopreservation of their embryos and to choose embryo donation to another couple if they perceived them as a child, whereas if they were more likely to see their embryo as a project, they were more likely to donate them to research (see also Chap. 8). The symbolic representation of gametes and embryos can therefore pose difficulties to parents when difficult decisions about their future need to be made. Targeted counselling may help make this easier.

Ethical Concerns About the Surrogate Arrangement

The important question posed decades ago concerned issues of reducing a surrogate mother to a 'vessel' or 'a baby making machine' (Krimmel 1983) rather than a person or a parent who relinquishes a child (McCormick 1987). The ethical issues relevant are numerous and mimic many of those described in Chap. 8 and these are represented in Fig. 9.3.

The Welfare of the Child

The welfare of the child is the most important concept debated because if no child or children were involved, the transaction would be taking place between consenting adults with no consequences to a minor. There is a distinct and understandable lack of evidence-based information about the welfare of any child born as a result of surrogacy. Any such children born within the United Kingdom are young and difficult to recruit for research purposes. Even in countries where surrogacy has been carried out for many more years than the United Kingdom, the children have not been studied sufficiently, with the exception of one research group in the United Kingdom (see Golombok and Tasker 2015). Secondly, the wellbeing of any children the surrogate already had has been of concern. The possibility that they may be affected by the separation of their (perceived) sibling could have unintended consequences for them. This can be particularly pertinent if the children knew money was involved in the creation, carrying and relinquishment of that gestational or partly genetic sibling. Thus, the moral and ethical argument rests on the potential for harm to the child, whether the surrogate child or the surrogate mothers' existing child(dren).

Procreative Liberty

All individuals have a right to produce children if they so wish. We (or laws and regulations) readily accept that in some cases medical assistance is required and much is therefore done to attempt to ensure that people's desires for a child are generally fulfilled. This includes creating families through non-traditional (and in some cases non-biological) means such as adoption, use of donated gametes and embryos. Surrogacy constitutes no more than an expansion of that continuum. The majority of the fertile population does not ask for assistance or permission to create a conceptus. One could therefore argue that any restrictions which are put upon those who need assistance could be interpreted as a breach of human rights and individual freedom. More importantly, why should infertile individuals be discriminated against having children? Within traditional conceptions, so much can and does go wrong in terms of the welfare of the child such as abuse, poor parenting, drug use, transmission of the AIDS virus, and so on, suggesting that imposing some sort of screening upon commissioning parents much like adoptive parents may not seem justifiable. All individuals also have a right to reproductive autonomy

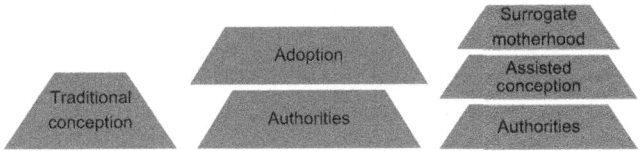

Fig. 9.4 Showing the increasing number of authorities responsible for bringing about and placing a surrogate baby, compared to adoption and traditional conceptions

and no law or regulation should be in a position to prevent individuals from exercising this right. However, the critical difference in traditional conception is that no other authority was involved in placing the child, whereas in adoption and surrogacy they are, as shown in Fig. 9.4.

Brazier et al. (1998, section 4.32) argue that procreation is not just a matter of individual freedom when it involves bringing another life into this world, 'whose welfare and autonomy deserve the highest attention from the state'. Of course, all procreation involves the potential bringing into being of a new dependent life but more people are involved in the bringing about a conception in surrogacy; therefore, the opportunity for conflict is greater than in non-assisted conceptions. It may therefore be useful to consider paying due attention to them. Also, because of the involvement of third parties, future claims of a genetic, gestational or social link can have an effect on the child and the adults involved. At present little evidence of detrimental effects on the adults or children involved in surrogacy is reported; it is nevertheless a moral issue that has not yet been put to the test. The debate should therefore continue and ideally it should include the voices of those involved. More research should be commissioned to urgently provide the evidence needed to inform practice.

Exploitation of the Surrogate

A surrogate arrangement can be perceived as a potentially exploitative arrangement since money is exchanged. However, if a surrogate mother is fully informed of the risks to herself the fact that payment is involved should be a secondary issue if she autonomously enters into the arrangement willing and aware—not exploited. It is possible however to retrospectively perceive an arrangement as exploitation. If a surrogate mother went into the arrangement fully informed but discovers upon delivery or following relinquishment of the baby that it was not the right thing for her to do, she may feel exploited because she has contractually and financially agreed to do this without understanding the emotional consequences of future feelings about the surrogate motherhood process. Thus, the risk of exploitation can be a possibility in surrogacy, be it an unpredictable risk. The unpredictable risk here is one of the regret of giving up a baby, not solely a risk of the payment involved. The issue of payment may then be irrelevant. Since in the United Kingdom surrogate motherhood arrangements are not enforceable in law, payment is not a reason to give up a baby. If a surrogate feels she cannot relinquish a baby, she does not have to do so. Thus, the argument shifts to one of emotional exploitation, not a financial one. One could argue that freedom of choice to procreate or not always carries a potential risk. The freedom could be interpreted as exploitation by the individual. For instance, the availability of contraceptive methods and delaying childbearing can, in retrospect, be a choice regretted if the individual had known in advance that parenting a child at a later age proves to be more difficult for that individual. Similarly, decisions to abort a foetus can be regretted retrospectively as the emotional risks of regret cannot be foreseen, despite a wealth of information available to those choosing these options. Not all decisions people make in reproductive terms are therefore necessarily interpreted as positive decisions at a later stage. This is not unique to surrogacy and the financial involvement can be irrelevant to the moral argument of exploitation.

Commodification

In moving away from the exploitation of the surrogate and considering the commodification of procreation, financial transactions are more relevant. There are cases of surrogacy where no financial transactions take place. This is usually when relatives or close friends agree to carry a baby for the infertile couple. Even here, some recompense for expenses such as travel to and from clinics will take place. In surrogacy where the couple were previously strangers, these same expenses constitute part of the payment. The remainder, regardless of the exact amount, tends to be paid for inconvenience, loss of earnings and to buy maternity clothes, vitamins and so on.

Commodification in surrogacy refers to 'baby selling'. It has been argued that within the arrangements taking place in the United Kingdom, or within the social 'contracts' some mention is made of the completion of the arrangement; for example, the full or last payment will be made upon relinquishment of the baby. Although this is not enforceable in law, it does portray the commodification, the buying and selling of a baby or of services provided to deliver a baby. Thus, if some form of payment was to be allowed constitutionally it would have to also accommodate provision for the paying commissioning couple, for example, some recompense to them if the child was not relinquished. This then constitutes commodification of a baby, not altruism. However, there are difficulties with such interpretations. If we consider the fact that many more infertile couples turn to clinics for treatment, payment for treatment to have a baby is also made. More explicitly, payment for donor gametes or embryos increases the cost. Is this any different from commodification of a baby in surrogacy? Baby buying and selling is therefore an immoral practice particularly if one considers the spin-off consequences, when the 'goods' obtained do not match expectations, those involved are treading a fine line. Furthermore, the recompense in surrogacy does not take place within a moral vacuum. It is a practice which was developed out of a need, just like payment for other forms of procreative assistance. It is also a practice which must be considered in light of a rapidly changing society and in light of a society which practices post-code lotteries and favouring the better-off against the worse-off within the population for other treatment opportunities. Within reproductive treatments, those with more money are in a better position to try the more expensive treatments and can do so more often, than those who are not financially as advantaged.

Admittedly, these other wrongs do not make baby selling right, but to see surrogacy in a moral vacuum is not appropriate or useful. Ultimately, whether we intuitively grimace at the fact that financial factors are relevant to much of what we do, they are undoubtedly important. Couples with children who split up will in some instances divide up the time each spends with their children according to the amount of financial support that is given to one by the other. The child support agency waives payment to the absent party if he or she spends more than a certain number of days and nights looking after the child(ren). Interestingly, Parker (1984) would argue that in surrogacy no babies are sold but the parental rights and responsibilities are being 'purchased', which has also been prohibited (Alexander and O'Driscoll 1980). The concern is that the traditional family would be destroyed by having money and profit introduced into the process of becoming a family. The long-term consequences of payment on the parents (surrogate and intended) as well as the child are not yet known but should not be underestimated. Guilt, regret and expectations of quality of purchase need to be considered in the longer term. However, there are two further points to raise. First, since there is nothing traditional about third-party-assisted conception families, particularly not where gestational or donor gametes are used in the arrangement (for a fuller discussion of non-traditional families see Part II, the chapter on separation and parenting, Chap. 6). Second, the destruction of the family occurs when the parents' feelings towards the child are 'contaminated' by the purchase of the child, with the child permanently damaged in these cases (Parker 1984, p. 32). According to Parker, these concerns are largely unfounded in practice and the 'black market' in parental rights does not apply to surrogate motherhood cases. Unfortunately, 30 years on and as was discussed in Chap. 8 and will be shown later, this is now in fact a very real concern in modern cross-border surrogate motherhood arrangements, with highly dubious individuals commissioning these babies for extremely unsavoury and illegal purposes.

Medical and Other Health Care Responsibilities

Offering treatment and helping people to achieve pregnancies in surrogate mothers, using IVF/ICSI or DI brings with it huge responsibilities for the welfare of all three in this reproductive ménage á trois. Lessons have been learned from experience that multiple pregnancies are generally one of the most significant harms to a mother (in this case the sur-

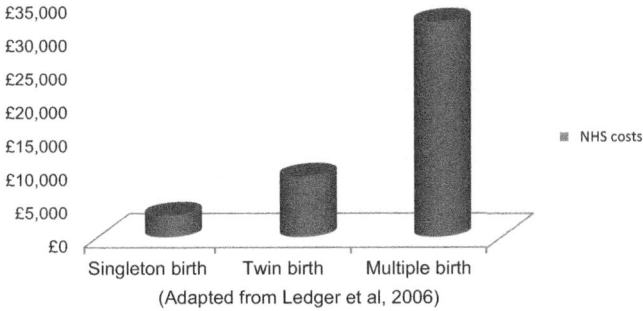

£35,000
£30,000
£25,000
£20,000
£15,000   ■ NHS costs
£10,000
£5,000
£0

Singleton birth   Twin birth   Multiple birth
(Adapted from Ledger et al, 2006)

Fig. 9.5 NHS costs of single, twin and multiple births (Adapted from Ledger et al. 2006)

rogate) and the baby. It also adds to substantial social and health risks and to medical costs to the NHS (Ledger et al. 2006). Figure 9.5 shows the total costs reported in 2006 in the United Kingdom (Ledger et al. 2006) per singleton birth, compared to twin and multiple births. The psychological costs are significantly higher too, as was confirmed in a recent meta-analysis comparing assisted conception twin/multiple births with naturally conceived twin/multiple births, and assisted conception twin/multiple births with assisted conception singleton births (van den Akker et al. 2016). Despite this, some countries still offer 'buy one get one free' (BOGOF) and other benefits of multiple pregnancies. Ethically this is unacceptable.

The increase in psychological morbidity and medical morbidity and mortality in medically assisted twins and multiple births, in tandem with increasing demands for these services, led to national and international efforts in developing legislation and guidelines (e.g. the HFEA, One at a Time 2006; Ledger et al. 2006; ASRM 2009; McLernon et al. 2010) to help improve the outcome to mothers and babies. Improvements in formal processes of reporting on current practices were also established (e.g. the IFFS; International Federation of Fertility Societies 1992). The IFFS reports data from 59 countries (Jones et al. 2011). The US Fertility Clinic Success Rate and Certification Act (1992) required clinics to report specific information on IVF cycles (Schieve et al. 1999). Improvements in IVF techniques led to lessening the need for multiple embryo transfers, as success rates with elective single embryo transfer (eSET) improved

(McLernon et al. 2010). These improved success rates are expected to eradicate the additional increases in maternal and infant morbidity and mortality. However, there were also unintended consequences of these monitoring regulations; it led to patients seeking the 'more successful' clinics and clinics, in turn, 'cherry picking' patients to improve their statistics (Brezina and Zhao 2012). Reporting therefore is useful and necessary, but also ethically problematic. The best clinics could charge more for their excellence in reported outcomes, leading to possible further inequalities between those seeking the services. In addition to that, the limitations set by some countries on numbers of embryos for transfer in any one cycle (Schieve et al. 1999; Hughes and DeJean 2010; Jones et al. 2011; Setti et al. 2011) and the rising costs also led to people travelling across the world (Collins and Cook 2010) to obtain better value for money IVF packages. This is even more pronounced when IVF surrogacy is used (Humbyrd 2009), demonstrating the ethics of cross-border surrogacy is far from resolved (Humbyrd 2009; Damelio and Sorenson 2008; van den Akker 2016).

In some countries, including the United Kingdom, all assisted conception pregnancies (including singleton) are considered 'high risk' pregnancies, since there is an increased rate of placenta and vasa praevia, and deep vein thrombosis is also more common in IVF mothers (Mounce 2006). It is likely therefore that gestational surrogate mothers too need to be treated as high risk ante-natally, although this would be possible in practice only if the surrogate mother reveals she is carrying a baby for someone else. Similarly, in the post-natal period, the mandatory midwifery and health visitor visits can give assisted conception mothers or surrogates due consideration only if they tell health care workers they had delivered an assisted conception pregnancy and/or surrogate baby.

The American College of Obstetricians and Gynecologists (ACOG, Grodin 1991) provides ethical statements, which it updates as and when new information becomes available. The 1990 statement updates the previously issued ethical statement on surrogate motherhood in 1983. In its 1990 statement, more autonomy and rights are given to the surrogate mother, as sole source of consent regarding events in pregnancy and postnatally, and gives her time in the post-partum to decide if she can proceed with the relinquishment of the surrogate-born baby. Interestingly, the 1990 ACOG statement also refers to the weightier link of gestation and birth than an intended parents' genetic link (Grodin 1991, p. 130.); only medical need for surrogacy is an accepted reason; payments in compensation to surrogates should be for her service in attempting to assist a commissioning couple, not for a successful outcome; and counselling and screening of both parties are advocated. A short consideration of social and psychological harm resulting from surrogacy is also noted such as the trivialisation of reproduction or regarding women and children as commodities and the surrogates' children fearing they too may be given away' respectively (Grodin 1991, p. 131).

Advances in perinatal medicine are also increasingly important in the perinatal care provision of surrogate mothers. Assuming surrogate mothers presenting for ante natal care during pregnancy reveal they are carrying a baby for a commissioning parent(s), medical and health care professionals have to deal with additional complex challenges. Foetal programming and therapy, epigenetics, foetal reduction and prenatal screening and diagnostic testing in a surrogate pregnancy may require an understanding of the contractual arrangement between the commissioning and surrogate parents. There could be a conflict of interest between the parties and these will need to be navigated wisely by the perinatal care team.

Monitoring, Record Keeping and Reporting

Burrell and Endozien (2014) report that many 'do it yourself' arrangements may be slipping under the radar particularly if arrangements are informally carried out between friends and family. In these cases of genetic surrogacy, the baby is handed over to the friend or family member without the involvement of the legal system. Moving house to hide the fact that no pregnancy took place in the intended couple is also likely so that subsequent medical help can be sought from the health care system as if this baby had been born into the commissioning couples' unit. In 2003, the HFEA reported a greater increase in surrogate births (of 10.2%) compared to the increase in IVF (of 8.3). Similarly, the annual rates of POs granted more than doubled between 2007 and 2011 (Crawshaw et al. 2012). Despite these increases, it is evident that many surrogate arrangements are not registered as surrogate births, and not all parents apply for POs, both from those coming into the United Kingdom with a baby commissioned abroad, and those from within the United Kingdom (Crawshaw et al. 2012).

The obvious legal problems are therefore exacerbated by the fact that surrogate motherhood is not considered 'a medical condition that has a rate or prevalence that can easily be tracked' (Reilly 2007, p. 483.), particularly not when genetic surrogacy is used. Instead, as previously noted, it has been described inappropriately as a social arrangement (Rivard and Hunter 2005). Despite this, some countries expect clinicians to report the number of embryos transferred (e.g. the United Kingdom, HFEA 2014). Other countries do not specify how many should be transferred (e.g. the United States, the Fertility Clinic Success Rate and Certification Act 1992), rendering all relevant authorities (medical, social and legal) to operate in a vacuum.

The Moral Position

Morally, the question, should reproductively compromised individuals remain childless, is a difficult and absurd one. Psychological distress is often unavoidable in individuals wanting to have children, and is reported for sub-fertile men (Glover et al. 1996; Daniluk 1988) and sub-fertile women (Wright et al. 1991; Slade et al. 2007). It is therefore not surprising that involuntarily childless couples go to great lengths to have a child. However, the long-term effects of surrogacy arrangements are not well documented. It is possible that the psychological distress of the uncertainties of surrogacy far exceeds the distress experienced by sub-fertile individuals. There is a moral obligation to support unusual reproduction effectively.

As noted in the previous section in this chapter, the incidence of surrogacy is impossible to estimate because of the informal arrangements which are known to take place (BMA 1996), and partial surrogacy does not necessarily require medical intervention. Proper statistics will need to be obtained on the incidence and types of surrogacy in the United

Kingdom and the differential effects these have on the intended and surrogate parents. Studies of recipient couples' viewpoints of DI have shown that expectations of success were unrealistic and knowledge of DI was insufficient (Bielawska-Batorowics 1994). It is therefore also essential that the information and advice given by voluntary agencies are accurate and comprehensive. Since we are no longer dealing with isolated cases of surrogacy, and this practice is likely to increase in the future, it is necessary to monitor the efficacy of information relayed by voluntary agencies, so that as far as is possible morally, the wider implications are catered for by the agents/clinics supporting the practices.

Quality of Information Received and Knowledge of Surrogacy

The legal literature on surrogacy is fraught with regulations which can be difficult to understand and can have devastating effects on all parties in the surrogacy agreement. The parties involved in surrogacy in the United Kingdom can never be sure of the child's future because the surrogacy arrangement cannot be enforced. It is unclear if information about paternity is known to the surrogate or the commissioning parents. For example, if either party reneges on the surrogacy agreement, the surrogate's husband/partner or if he does not exist, the intended father (if biological) could be held responsible as the legal father of the child. At 18 the child can obtain a certified copy of the original birth record which includes the surrogate mother's name. Little information about the surrogate mothers' knowledge of legislation exists. In 2003 van den Akker reported that the majority of surrogates felt well informed about the practical and medical aspects of surrogacy, but knowledge of the legal, psychological and social aspects was not adequate. Their primary sources of information were from the agency, the clinic or their own research, yet most surrogate mothers became pregnant within 2–12 months of knowing the commissioning couple. However, it is important that they fully understand all the possible consequences. The child, for example, is in a position to claim damages for negligence against the surrogate or intended parents if they failed to seek appropriate testing for infectious diseases and the child is born with a disease (Human Fertilisation and Embryology Act 1990). There is an abundance of evidence from the field of screening and diagnostic tests in pregnancy of the effects of uncertainty on outcome leading to increased anxiety (for a review, see van den Akker 1993). The uncertainty of outcomes in surrogacy might have equally detrimental effects. This therefore suggests the uncertainty and anxiety associated with knowledge of surrogacy regulations will need to be assessed.

The Human Rights Perspective

Because of the historical oppression of women, the Convention of the Elimination of Discrimination Against Women (CEDAW 1979), is concerned specifically with issues arising in relation to maternity (see Article 4). The international private law protects the rights of women in relation to paid employment maternity leave (Holtmaat and Naber 2011). No rights of maternity as work or the occupation of maternity exist, because at the time no such occupations were spoken of and IVF surrogacy did not yet (officially) exist. Young (1990) explains CEDAW's stance on the discrimination against women as a form of oppression which can be five-fold as shown in Fig. 9.6.

According to CEDAW, all women have a right to choose their work and to choose to reproduce or not. However, surrogacy is not a traditional form of employment, and surrogate mothers do not have full control over their surrogate reproduction in a traditional sense. One can argue two ways about the employment: first, that it should not be considered employment as the creation and nurturing of an unborn child is at stake and, second, that the 'employees' are exploited by the companies/organisations/clinics who 'employ' them. Similarly, especially in cross-border surrogacy, one could argue that the surrogate mother has in effect little control over the conception, the pregnancy or the delivery, as most of these are decided and controlled by the brokering companies or clinics. If viewed this way, the five oppression concepts outlined by Young (1990) above could be argued to apply to cross-border surrogates, as these are nearly always extremely poor and lack any form of education.

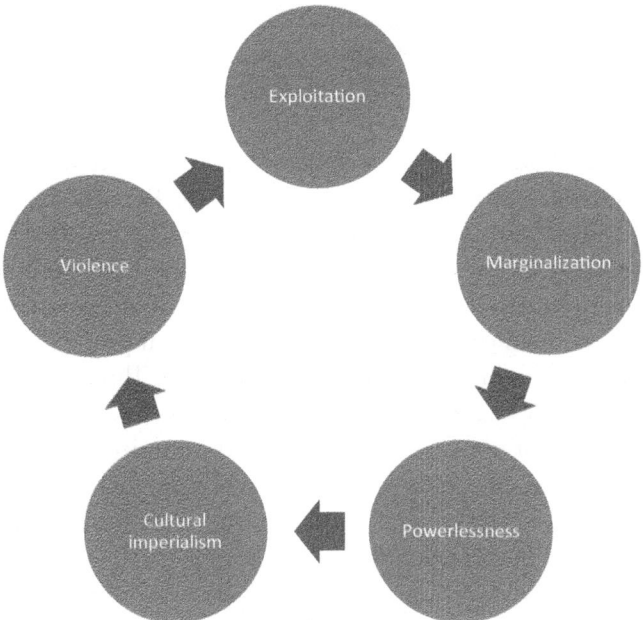

Fig. 9.6 Showing the five areas of oppression as identified by Young (1990)

In poor resource areas, violence against women is common as there is no accessible law protecting them, powerlessness and marginalisation is common in some cultures and cultural imperialism applies to the wealthier commissioning couples from the Western world commissioning surrogates who are poor and live in societies where poverty and inequalities are prolific. Surrogates should therefore be particularly protected from potential exploitation, because it is morally wrong to exploit vulnerable others. Protection is necessary despite the fact that some of the poor and potentially vulnerable surrogates in India are reported to believe that the work of surrogacy is an occupation they themselves are happy with (Karandikar et al. 2014). The argument that women should have the right to seek employment in any way they wish and if they choose to use their reproductive organs for this purpose this should be their right (Mason 1998) is acceptable only if this is done from an empowered perspective, not a vulnerable one. It is highly questionable if these surrogate mothers go into the arrangement empowered since they have few options in life, have no money, no education and few other prospects. Finally, there is the issue of commodification of the child that may not be within anyone's right, including that of a surrogate mother, and which is considered immoral.

There are also 'add on' effects of surrogacy arrangements in countries where the surrogate mothers are extremely poor. Reports demonstrating social shunning of surrogates by their community (Lycett 2009), receiving judgement about carrying 'another' man's child, or that of a 'homosexual' man, and being accused of adultery (Hochschild 2009) have been emerging. Neither the medical nor legal informed consent processes adequately incorporate these potential harms to the surrogate. Thus the psychosocial and emotional short- and longer-term effects on the surrogate, her own children and that of her family within the community in which she lives are not yet considered appropriate risks. Ethically, this is an inexcusable oversight, as the physical risks should not outweigh the mental health of the surrogate and those dependent upon her.

The Universal Declaration of Human Rights (UDHR; Article 16.1), drafted over 50 years ago, notes that adult men and women have the right to marry and build a family. The 'family' has not been defined. Since modern families deviate from traditional families, and the declaration has not been updated to include contemporary family forms, or those using unusual reproduction, the issue of rights and families remain undefined, and may therefore be assumed to remain as stated. Commissioning parents are therefore also potentially vulnerable, because traditionalists may not recognise their needs and apparent equal rights to found a family. Commissioning parents are better protected than surrogate mothers and they are in a financially better position to seek legal advice should that be necessary. Despite this, they too can be exploited by the brokering agents or intermediary services or by the surrogate mothers directly if no intermediaries are used. Commissioning couples pay relatively large sums of money to have surrogate babies. These emotionally vulnerable people can be exploited, more money can be asked for or the baby can be withheld from them if a surrogate reneges on the arrangement. Again, the legality of these issues depends upon the country where the surrogacy takes place. In the United Kingdom, for example, the mother giving birth is the automatic birth mother (although judges have overruled this), whereas in

India this is not the case. Despite this, there is no definitive human right to a child and no definitive human right to parent a child (Stark 2012).

The United Nations Convention on the Rights of the Child (UN CRC) 1989, which has been ratified by all members of the UN except Somalia and the United States, specifies that the rights of children and their best interests should be afforded primary consideration. The CRC (Article 6) also refers to the right of the child to identity, name and family relationships, and (in Article 7) the right to know and be cared for by their parents. More importantly from a surrogate child's perspective, Article 9 refers to the child's right not to be separated from their parents without judicial review, Article 7 stipulates the child's right not to be stateless and Article 8 refers to the child's right that their nationality is to be guaranteed. In practice these rights may be difficult to guarantee in (particularly cross-border) surrogacy arrangements.

The British Association for Social Workers (BASW 2016) promotes best practices to secure the welfare of others, and upholds the Universal Declaration of Human Rights and other UN declarations. Regarding surrogacy, both sets of adults (surrogate and commissioning) and the child's welfare across the lifespan are paramount considerations for social workers. BASW recognises a number of issues relating to surrogacy, and puts the best interests of the child at the top of their priorities list in surrogacy arrangements (see BASW Code of Ethics at 2.1, Ethical Practice Principles 3, 7, 8). This paramountcy is recognised in part in UK legislation concerning the transfer of legal parenthood following surrogacy through the making of PO (Section 54, HFEAAct 2008) and The Human Fertilisation and Embryology (POs) (HFEA Regulations 2010) on the welfare of surrogate-born children. This way, regardless of the potential complications arising from a surrogacy arrangement, the welfare of the child is at the forefront.

Welfare of the Child

In commercial surrogacy, the interests tend to be that of the parents' desires to be parents rather than the needs of child(ren) born through the arrangements. Foetuses may be at risk of disabilities or suffering poor health (e.g. through multiple births) babies may not attach to the eventual parents in a timely manner. Surrogate-born children may learn that their birth mother was at risk of receiving poor ante- and post-natal care, particularly in cross-border arrangements. International law fails to consider poor practice in cross-border surrogacy (van den Akker 2016), and monitoring such international movement for services is virtually impossible. However, ethically and morally there is good reason why monitoring cross-border surrogate activity is important. Surrogate children are, for example, deprived of their surrogate mother's bond which would have had time to develop during the pregnancy. We do not yet know what the effects of this are likely to be, or if newborn babies are resilient in their adaptation to different caregivers. Unlike in adoption, the child is created specifically for the intended parents' welfare, whereas in adoption the child is usually non-intentionally created and needs to part with its birth mother for its welfare. Some surrogate-born children may develop a 'disconnect' of where they belong (Cuthbert and Fronek 2014, p. 7), others may develop issues with identity as is reported in, for example, some donor-conceived adults (van den Akker et al. 2015), although it is expected that the majority are likely to do well.

As noted above, surrogate-born children have the right under Article 7 of the CRC to 'know' their parents. How they themselves come to define 'parents' may depend upon where in the world they are brought up. This may be more complex in children conceived via a surrogate arrangement, because they may not know the identity of the person who gave birth to them, any gamete or embryo donors, or even the commissioning parents (if they failed to honour their agreement). Surrogate-born children are therefore in a unique situation of potentially having several, or no known, 'parents'. Even in gestational surrogacy where the surrogate does not use her own oocyte it is recognised that there is a biological relationship which can influence the health and wellbeing of the foetus. The gestational surrogate may also contribute some of the physiological predispositions whereas the genetic surrogate will contribute additional genetic imprints onto the new baby (Egliston et al. 2007; Ombelet et al. 2005). If donor gametes were used, the child will inherit genetic material from these donors. There are few jurisdictions where treatment centres or gamete donor banks keep registries that enable a surrogate-born child to trace the identity of all their 'parents'—however, this is useful to the child only if their commissioning parents make them aware of their origins.

Surrogate-born children therefore do not necessarily find out what their true identity and the identity of all their 'parents' is, not even via birth registration systems. For example, anyone born in the United Kingdom, but conceived using donor gametes, will not find that recorded on the birth certificate nor is there a system anywhere (including within the birth registry) that enables such information to be made available to them despite their legal right to access their donor's identity (if they were born after April 2005 and conceived in a UK licensed centre). Again, being made aware of their third-party conception is necessary in the first place, which is not always the case (van den Akker 2012; Tallandini et al. 2016). Even when a child is born in another country, a child may not be legally registered with the parents if they wish to circumvent the child's surrogate birth despite UK regulations stipulating legal parenthood post surrogacy must be obtained by declaring their status and applying for parental orders. For a commissioning parent not to be recognised as the legal mother of the surrogate baby can have implications for the surrogate mother if she remains registered as the legal mother of the child. For example, succession rights, pension and taxation are then tied to the surrogate mother, not the intended mother. Here a child born to a surrogate would have the rights to a share of the inheritance, but not that of the intended mother as was shown in MR & Anor v An Chlaraitheoir and Others (2013).

Of course, denying a child a nationality is not done intentionally, but done to protect the parents' secrecy about the child's conception. There are cases where a child was not able to leave the country where they were born because the parents failed to notify the appropriate authorities. The likelihood is that either of the parents travels home and leaves the child with the remaining commissioning parent (or the surrogate) until a solution to bring the child to its intended home is found. This may have implications for attachment bonds the child is not developing with that absent parent as was discussed in Chap. 6. Arranging a pre-birth PO or similar could solve that problem, but that goes at the expense of the child knowing the true origins of its birth and potentially against the welfare of the surrogate mother. According to Crawshaw (2016) a system of tracking would be compatible with the best interests of the child and

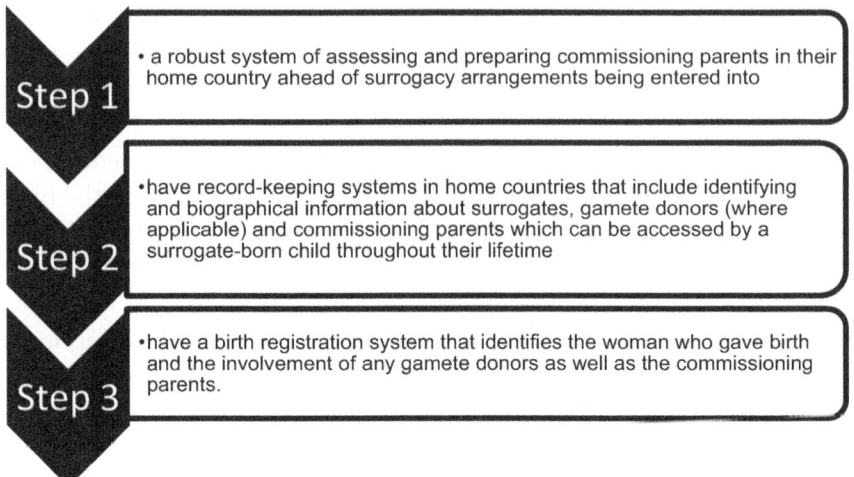

Step 1
• a robust system of assessing and preparing commissioning parents in their home country ahead of surrogacy arrangements being entered into

Step 2
•have record-keeping systems in home countries that include identifying and biographical information about surrogates, gamete donors (where applicable) and commissioning parents which can be accessed by a surrogate-born child throughout their lifetime

Step 3
•have a birth registration system that identifies the woman who gave birth and the involvement of any gamete donors as well as the commissioning parents.

Fig. 9.7 A monitoring and registration system for surrogate births

would enable nationality issues to be resolved (e.g. MR & Anor v An Chlaraitheoir and Others (2013) IEHC); see Fig. 9.7.

The Hague Convention on Protection of Children and Co-operation in Respect of Intercountry Adoption (HCCH 1993) has been signed by over 80 states who all agree upon the core principles of upholding the welfare of the child in inter-country adoptions. The Hague Permanent Bureau is currently seeking to determine if there is a need to draw up international private law on cross-border surrogacy (Hague Conference on International Private Law 2012). In 2015, the Council on General Affairs and Policy of the Hague Conference decided that an Experts' Group should be convened to explore this. The Council decided that an Experts' Group would meet and report to the 2016 Council meeting (HCCH 2016).

Welfare of the Surrogate

Surrogate mothers are exposed to risks they would not have incurred if they had not become a surrogate mother. The risks include medical, social and psychological risks, and they are not effectively controlled for in legislation or in (clinical) practice and are therefore ethically problematic

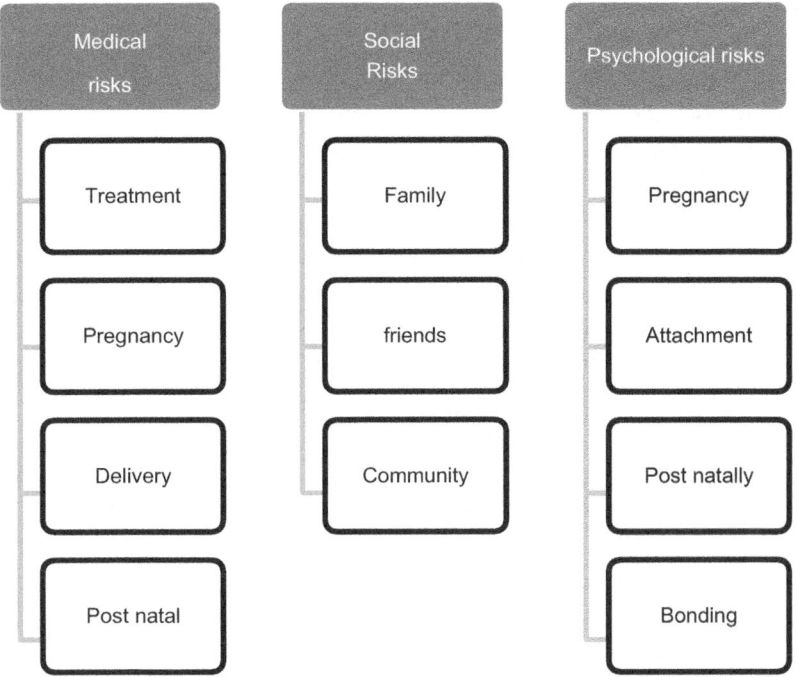

Fig. 9.8 Some of the medical, social and psychological risks identified

consequences of surrogate motherhood arrangements. Each of these risks is discussed in the previous chapters in Parts I and II as relevant to the surrogate mother, the commissioning couple or the surrogate-born individual, and they are also briefly described here as they are relevant to legal and ethical issues. The risks can be quantified in the following terms; see Fig. 9.8.

The medical risks associated with the treatment include adverse reactions to the hormonal treatment for IVF. The medical risks associated with the pregnancy include the same unplanned and unanticipated pregnancy and delivery complications which can affect any pregnancy or post-partum period. Post-natally, the mother can continue to have psychological or medical complications in the immediate and the long term, such as post-natal depression or post-natal psychosis, reduced muscular strength or incontinence. However, in the case of surrogate mothers, unlike their own conceptions, these risks are taken on behalf of someone else who will not subject themselves to these risks, but instead enjoy the result of the risk undertaken by the surrogate mother. There are also ethical concerns about the incidence of planned and potentially anticipated risks of medical procedures including multiple births, twinning rates and C-sections performed on some surrogates as a matter of course, instead of as a matter of medical indications. Twins and multiple births are a risk to maternal and infant health and wellbeing with morbidity and mortality much increased (van den Akker 2016). The welfare of premature and low birth weight infants is compromised, and since this is an intervention, the responsibility for the child's welfare lies with the clinicians involved in bringing multiple births about. In Indian surrogates, there is evidence that multiple embryos are regularly transferred (SAMA-Resource group for Women and Children 2012) and that they routinely undergo C-sections in some clinics (see, e.g. Carney 2010).

It is uncertain if all surrogates routinely gave informed consent to these or other procedures (Palattiyil et al. 2010, p. 695). Carney (2010) reported on an Indian surrogate who died following a haemorrhage after a surrogate baby was delivered. This begs the ethical question if the surrogates enter these arrangements knowingly, or if they have failed to understand the risks associated with pregnancy and delivery It is likely that they do not. van den Akker's 2003 British surrogates who were all literate, lacked appropriate knowledge of the implications of the medical procedures involved as was mentioned in the previous section. Any woman who has had a previous C-section must be monitored and may be more likely to need a C-section in a new pregnancy, and if these women are poor it is unlikely they can afford to seek appropriate medical care in their own subsequent pregnancies, putting them at a new and increased risk they may not have been aware of when they consented to be a surrogate. In fact Fasouliotis and Schenker (1999) note that anyone who lives in extreme poverty and who has limited education cannot provide informed consent.

The social risks associated with the surrogate arrangement concern reports discussed in more detail in previous chapters of fears that the surrogate's own children may think they could be given away or sold, family may judge their actions and grandparents may find it hard to cope with not being able to grandparent a genetically linked surrogate child. The surrogates' friends and community may judge the surrogate mother (Hochschild 2009; Pande 2009), resulting in social stigma, lack of social support and isolation. Ethically, to practice surrogacy under these conditions in countries where stigma can ruin lives and cause social instability and psychological hardship cannot be ignored or condoned. The psychological risks of the medical treatment, the pregnancy and the delivery are also numerous. Many pregnant women experience somewhat higher anxiety than if they were not pregnant, which may continue into the post-natal period. Additionally, the surrogate mother may attach to the foetus in pregnancy which can lead to anxiety about having to relinquish the baby or depression once the baby has been born. The psychological risks of attachment and bonding to the baby upon, and following, the delivery are also problematic, and discussed in detail in Chap. 6 on separation and parenting. Surrogate mothers may also experience post-natal depression or psychosis following the pregnancy and birth, which is rarely mentioned in the research literature of surrogate mothers (van den Akker 2003). These risks must be acknowledged and require ethical scrutiny perhaps feeding into new legislation to ensure the practice becomes ethically sound.

Summary

Ethical, moral and human rights issues penetrate every aspect of surrogate motherhood practices, and human rights have been shown to be violated. The dilemmas are difficult to reconcile but that should not pose a deterrent to achieving a workable and sound framework for surrogate motherhood. Surrogacy is increasing across the world, can be exploitative (particularly for babies) and prey upon the vulnerability of commissioning couples and surrogates alike. It therefore needs continuing ethical and moral scrutiny as described in this chapter and it needs regulating at some level (see next Chap. 10). The parties involved in surrogacy mostly enter the agreement to fulfil a need, a need for the commissioning parents to build a family and a belief for the surrogate mother to help them achieve this goal and/or a need to be compensated for their efforts to bear and deliver a baby for the commissioning couple. We may not have to wait for future generations to question the relatively lame approach researchers, practitioners and policymakers have adapted to current questionable surrogate motherhood practices. History will no doubt judge the moral, ethical and human rights consequences of the commissioning parent and surrogate motherhood practices, as well as the professionals' (in)actions.

References

Alexander, L. A., & O'Driscoll, L. H. (1980). Stork markets: An analysis of "baby selling". *Journal of Libertarian Studies, 4*(2), 173–196.

Andrews, L. B. (1987). The aftermath of Baby M: Proposed state laws on surrogate motherhood. *Hastings Center Report, 17*(5), 31–40.

Andrews, L. B., & Elster, N. (2000). Regulating reproductive technologies. *Journal of Legal Medicine, 21*, 35–65.

Aramesh, K. (2009). Iran's experience with surrogate motherhood: An Islamic view and ethical concerns. *Journal of Medical Ethics, 35*, 320–322.

ASRM. (2009). Practice Committee, guidelines on number of embryos transferred. *Fertility & Sterility, 95*(5), 1518–1519.

Baier, A. (1986). Trust and antitrust. *Ethics, 96*, 231–260.

Baier, A. (1994). *Moral prejudices: Essays on ethics*. Cambridge: Harvard University Press.

Banjerjee, S., & Basu, S. (2009). Rent a womb: Surrogate selection, investment incentives and contracting. *Journal of Economic & Behavioural Organisation, 69*(3), 260–273.

Baykal, B., Korkmaz, C., Ceyhan, S., Goktolga, U., & Baser, I. (2008). Opinions of infertile Turkish women on gamete donation and gestational surrogacy. *Fertility & Sterility, 89*(4), 817–822.

Beauchamp, T., & Childless, J. F. (1994). *Principles of biomedical ethics* (4th ed.). New York: Oxford University Press.

Bielawska-batorowicz, E. (1994). Artificial insemination by donor – An investigation of recipient couples' viewpoints. *Journal of Reproductive and Infant Psychology, 12*(2), 123–126.

Brazier, M., Campbell, A., & Golombok, S. (1998). *Surrogacy review for health ministers of current arrangements for payments and regulation*. Report of the review team. Cm 4068. London: Department of Health.

Brezina, P. R., & Zhao, Y. (2012). Review article: The ethical, legal and social issues impacted by modern assisted reproductive technologies. *Obstetrics and Gynecology International, 2012*, 1–7.

British Association of Social Workers (BASW). (2016). *Code of practice*, 11–18. http://cdn.basw.co.uk/upload/basw_112315-7.pdf. Accessed 19 Oct 2016.

British Medical Association (BMA). (1996). *Changing conceptions of motherhood. The practice of surrogacy in Britain.* London: BMA.

Bromfield, N., & Rotabi, K. (2014). Global surrogacy, exploitation, human rights and international private law: A pragmatic stance and policy recommendations. *Global Social Welfare, 1*(3), 123–135.

Bruno, C., Dudkiewicz-Sibony, C., Berthaut, I., Weil, E., Brunet, L., Fortier, C., Pfeffer, J., Ravel, C., Fauque, P., Mathieu, E., Antoine, J. M., Kotti, S., & Mandelbaum, J. (2016). Survey of 243 ART patients having made a final disposition decision about their surplus cryopreserved embryos: The crucial role of symbolic embryo representation. *Human Reproduction, 31*(7), 1508–1514.

Burrell, C., & Endozien, L. C. (2014). Surrogacy in modern obstetric practice. *Seminars in Fetal and Neonatal Medicine, 19*, 272–278.

Carney, S. (2010). Mother Jones: Inside India's rent-a-womb business. http://www.motherjones.com/politics/2010/02/surrogacy-tourism-india-nayna-patel. Accessed 19 Oct 2016.

Charo, R. A. (1988). Legislative approaches to surrogate motherhood. *Law, Medicine and Health Care, 16*(1–2), 96–112.

Collins, J., & Cook, J. (2010). Cross border reproductive care: Now and into the future. *Fertility & Sterility, 94*(1), e25–e26.

Convention of the Elimination of Discrimination Against Women (CEDAW). (1979). United Nations entity for gender equality and the empowerment of women. http://www.un.org/womenwatch/daw/cedaw/

Crawshaw, M. (2016). 13th programme consultation – PROGAR RESPONSE – 30th September 2016. http://cdn.basw.co.uk/upload/basw_40000-5.pdf. Accessed 19 Oct 2016.

Crawshaw, M., Blyth, E., & van den Akker, O. (2012). The changing profile of surrogacy in the UK – Implications for policy and practice. *Journal of Social Welfare and Family Law, 34*, 1–11.

Cuthbert, D., & Fronek, P. (2014). Perfecting adoption? Reflections on the rise of commercial offshore surrogacy and family formation in Australia. In D. Higgins (Ed.), *Children and families in Australia: Selected policy, legal and practical issues.* Melbourne: Australian Institute of Family Studies.

Damelio, J., & Sorensen, K. (2008). Enhancing autonomy in paid surrogacy. *Bioethics, 22*(5), 269–277.

Daniluk, J. (1988). Infertility: Intrapersonal and interpersonal impact. *Fertility & Sterility, 49*(6), 982–990.

de Lacey, S. (2005). Parent identity and "virtual" children: Why patients discard rather than donate unused embryos. *Human Reproduction, 20*, 1661–1669.

Donchin, A. (2010). Reproductive tourism and the quest for global gender justice. *Bioethics, 24*(7), 323–332.

Egliston, K. A., McMahon, C., & Austin, M. (2007). Stress in pregnancy and infant HPA axis function: Conceptual and methodological issues relating to the use of salivary cortisol as an outcome measure. *Psychoneuroendocrinology, 32*, 1–13.

Elias, S., & Annas, G. J. (1986). Social policy considerations in non-coital reproduction. *Journal of the American Medical Association, 255*, 62–68.

Erin, C. A., & Harris, J. (1991). Surrogacy. *Bailliere's Best Practice in Research & Clinical Obstetrics & Gynaecology, 5*(3), 611–635.

Erlen, J. A., & Holzman, I. R. (1990). Evolving issues in surrogate motherhood. *Health Care for Women International, 11*, 319–329.

Fasouliotis, S., & Schenker, J. (1999). Social aspects of assisted reproduction. *Human Reproduction Update, 5*(1), 26–39.

Foucault, M. (1980). In C. Gordon (Ed.), *Power/knowledge: Selected interviews and other writings 1972–1977*. New York: Pantheon.

Galloway, K. (2015). Theoretical approaches to human dignity, human rights and surrogacy. In P. Gerber & K. O'Byrne (Eds.), *Surrogacy, law and human rights* (pp. 13–29). Burlington: Ashgate.

Gamble, N. (2009). Crossing the line: The legal and ethical problems of foreign surrogacy. *Reproductive Biomedicine Online, 19*(2), 151–152.

Giddens, A. (1984). *The constitution of society: Outline of the theory of structuration*. Berkeley/Los Angeles: University of California Press.

Glover, L., Gannon, K., Sherr, L., & Abel, P. (1996). Distress in sub-fertile men: A longitudinal study. *Journal of Reproductive and Infant Psychology, 14*, 23–36.

Golombok, S., & Tasker, F. (2015). Socio-emotional development in changing families. In M. E. Lamb (Vol. Ed.) & R. M. Lerner (Series Ed.), *Handbook of child psychology and developmental science* (7th ed, Vol. 3, pp. 419–463), Social, emotional and personality development. Hoboken: Wiley.

Grodin, M. A. (1991). Ethical issues in surrogate motherhood. Statement. *Women's Health Institute, 1*(3), 129–134.

Hadfield, G. (1995). The dilemma of choice: A feminist perspective on the limits of freedom of contract. *Osgoode Hall Law Journal, 33*, 337–351.

HCCH. (1993). 33: Convention of 29 May 1993 on protection of children and co-operation in respect of intercountry adoption. https://www.hcch.net/en/instruments/conventions/full-text/?cid=69. Accessed 19 Oct 2016.

HCCH. (2016). Hague conference on private international law. The parentage/surrogacy project. The private international law issues surrounding the status of children, including issues arising from international surrogacy arrangements. https://www.hcch.net/en/projects/legislative-projects/parentage-surrogacy. Accessed 19 Oct 2016.

HCCH Hague Conference on Private International Law. (2012). Overview on the world organisation for cross-border co-operation in civil and commercial matters. http://www.hcch.net/index_en.pho?act=text.displaye&tid=26. (Not working????) Accessed 22 May 2016.

Healy, J. M. (1984). Legal regulation of artificial insemination and the new reproductive techniques – The search for clarification continues. In A. Milunsky & G. J. Annas (Eds.), *Genetics and the law III* (pp. 139–145). Boston: Plenum.

Held, V. (1993). *Feminist morality: Transforming culture, society and politics.* Chicago: The University of Chicago Press.

HFEA. (2003). *Code of practice*, 6th edn. http://www.hfea.gov.uk/docs/Code_of_Practice_Sixth_Edition.pdf. Accessed 17 Oct 2016.

HFEA. (2008). HFEA Act, Section 54. http://www.legislation.gov.uk/uksi/2009/2232/made. Accessed 19 Oct 2016.

HFEA. (2010). HFEA regulations. http://www.hfea.gov.uk/501.html. Accessed 19 Oct 2016.

HFEA. (2014). HFEA releases third report on fertility trends. http://www.hfea.gov.uk/8828.html. Accessed 17 Oct 2016.

HFEA 'One at a Time'. (2006). Report. http://www.hfea.gov.uk/Multiple-births-after-IVF.html. Accessed 17 Oct 2016.

Hochschild, A. (2009). Childbirth at the global crossroads. *The American prospect.* http://prospect.org/article/childbirth-global-crossroads-0. Accessed 19 Oct 2016.

Holtmaat, R., & Naber, J. (2011). *Women's human rights and culture: From deadlock to dialogue.* Leiden: Intersentia. ISBN 978-94-000-0137-4.

Hope, T., Savulescu, J., & Hendrick, J. (2003). *Medical ethics and law: The core curriculum.* London: Livingstone.

Hughes, E. G., & DeJean, D. (2010). Cross border fertility services in North America: A survey of Canadian and American providers. *Fertility & Sterility, 94*(1), e16–e19.

Human Fertilisation and Embryology Act. (1990). Legislation.Gov.UK. National Archives. http://www.legislation.gov.uk/ukpga/1990/37/contents

Humbyrd, C. (2009). Fair trade international surrogacy. *Developing World Bioethics, 9*(3), 111–118.

Jin, X., Wang, G., Liu, S., Liu, M., Zhang, J. And Shi, Y. (2013) Patients' attitudes toward the surplus frozen embryos in China. *Biomedical Research International,* 934567, 8 pages. doi:10.1155/2013/934567.

Johnson, I. (1999). Regulation of assisted reproductive technology: The Australian experience. In P. R. Brinsden (Ed.), *A textbook of in vitro fertilization and assisted reproduction* (pp. 424–427). Carnforth/New York: Parthenon.

Jones, H. W., Cooke, I., Kempers, R., Brinsden, P., & Saunders, D. (2011). International Federation of Fertility Societies surveillance. *Fertility & Sterility, 95*(2), 491.

Karandikar, S., et al. (2014). Economic necessity or Noble cause? A qualitative study exploring motivations for gestational surrogacy in Gujarat, India. *Affilia: Journal of Women and Social Work, 165,* 23.

Krimmel, H. T. (1983). The case against surrogate parenting. *Hastings Center Report, 13*(5), 35–39.

Ledger, W. L., Anumba, D., Marlow, N., Thomas, C. M., & Wilson, E. C. (2006). The costs to the NHS of multiple births after IVF treatment in the UK. *Journal of Obstetrics and Gynaecology, 113*(1), 21–25.

Leeton, J. (1991). The current status of IVF surrogacy in Australia. *Australian & New Zealand Journal of Obstetrics & Gynecology, 31,* 260–262.

Lycett, E. (2009). Surrogacy. In G. Bentley & R. Mace (Eds.), *Substitute parents: Biological and social perspectives on alloparenting in human societies.* New York: Berghahn Books.

Lyerly, A. D., Steinhauser, K., Namey, E., Tulsky, J., Cook-Deegan, R., Sugarman, J., Walmer, D., Faden, R., & Wallach, E. (2006). Factors that affect infertility patients' decisions about disposition of frozen embryos. *Fertility & Sterility, 85,* 1623–1630.

Mason, M. A. (1998). The modern American stepfamily: Problems and possibilities. In M. A. Mason, A. Skolnick, & S. D. Sugarman (Eds.), *All our families* (pp. 95–116). New York: Oxford University Press.

McCormick, R. A. (1987). Surrogate motherhood: A stillborn idea. *Second Opinion, 5,* 128–132.

McLernon, D. J., Harrild, K., Bergh, C., et al. (2010). Clinical effectiveness of single versus double embryo transfer: Meta-analysis of individual patient data from randomised trials. *British Medical Journal, 341,* c6954.

Mohler-Kuo, M., Zellweger, U., Duran, A., Hohl, M. K., Gutzwiller, F., & Mutsch, M. (2009). Attitudes of couples towards the destination of surplus embryos: Results among couples with cryopreserved embryos in Switzerland. *Human Reproduction, 24,* 1930–1938.

Mounce, G. (2006). Assisted reproduction: What do midwives need to know? The Royal College of Midwives. https://www.rcm.org.uk/learning-and-career/learning-and-research/in-depth-papers/assisted-reproduction-what-do-midwives

MR & Anor v An Chlaraitheoir and Others (2013) IEHC. http://www.bailii.org/ie/cases/IEHC/2013/H91.html. Accessed 19 Oct 2016.

Nachtigall, R. D., MacDougall, K., Harrington, J., Duff, J., Lee, M., & Becker, G. (2009). How couples who have undergone in-vitro-fertilization decide what to do with surplus frozen embryos. *Fertility and Sterility, 92,* 2094–2096.

Ombelet, W., De Sutter, P., Van der Elst, J., & Martens, G. (2005). Multiple gestation and infertility treatment: Registration, reflection and reaction – The Belgian project. *Human Reproduction Update, 11,* 3–14.

Oxman, R. B. (1993). California's experiment in surrogacy. *Lancet, 341,* 1468–1469.

Palattiyil, G., Blyth, E., Sidhva, D., & Balakrishnan, G. (2010). Globalization and cross-border reproductive services: Ethical implications of surrogacy in India for social work. *International Social Work, 53*(5), 686–700.

Pande, A. (2009). 'Not an angel' not 'a whore': Surrogates as 'dirty workers' in India. *Indian Journal of Gender Studies, 16,* 141–173.

Parker, P. J. (1984). Surrogate motherhood, psychiatric screening and informed consent, baby selling, and public policy. *Bulletin of the American Academy of Psychiatry Law, 12*(1), 21–39.

Pateman, C. (1988). *The sexual contract.* Stanford: Stanford University Press.

Provoost, V., Pennings, G., DeSutter, P., van de Velde, A., De Lissnyder, E., & Dhont, M. (2009). Infertility patients' beliefs about their embryos and their disposition preferences. *Human Reproduction, 24,* 896–905.

Provoost, V., Pennings, G., DeSutter, P., van de Velde, A., & Dhont, M. (2012). Trends in embryo disposition decisions: Patients' responses to a 15-year mailing program. *Human Reproduction, 27,* 506–514.

Purewal, S., & van den Akker, O. B. A. (2007). The socio-cultural and biological meaning of parenthood. *Journal of Psychosomatic Obstetrics and Gynaecology, 28*(3), 79–86.

Radin, M. J. (1987). Market-inalienability. *Harvard Law Review, 100*(8), 1849–1937.

Ramskold, L. H., & Posner, M. P. (2013). Commercial surrogacy: How provisions of monetary remuneration and powers of international law can prevent exploitation of gestational surrogates. *Journal of Medical Ethics, 39*, 397–402.

Reilly, D. R. (2007). Surrogate pregnancy: A guide for Canadian prenatal health care providers. *Canadian Medical Association Journal, 176*(4), 483–485.

Rimm, J. (2009). Booming baby business: Regulating commercial surrogacy in India. *University of Pennsylvania Journal of International Law, 30*(4), 1429–1462.

Rivard, G., & Hunter, J. (2005). *The law of assisted human reproduction.* Markham: LexisNexis Canada, Inc.

Rothenberg, K. H. (1988). Baby M, the surrogacy contract, and the health care professional: Unanswered questions. *Law Medicine & Health Care, 16*, 113–120.

SAMA Resource Group for Women and Health. (2012). Birthing a market: A study on commercial surrogacy. http://www.samawomenshealth.org/research/inter-and-intra-south-dialogues-capacity-building-and-advocacy-assisted-reproductive

Samorinha, C., Pereira, M., Machado, H., Figueiredo, B., & Silba, S. (2014). Factors associated with the donation and non-donation of embryos for research: A systematic review. *Human Reproduction Update, 20*, 641–655.

Saravanan, S. (2013). An ethnomethodological approach to examine exploitation in the context of capacity, trust and experience of commercial surrogacy in India. *Philosophy, Ethics and Humanities, 8*(10), 1–12.

Schieve, L. A., Peterson, H. B., Meilke, S. F., et al. (1999). Live birth rates and multiple birth risk using in vitro fertilization. *Journal of the American Medical Association, 282*(19), 1832–1838.

Setti, P. E. L., Albani, E., Cesana, A., et al. (2011). Italian constitutional court modifications of a restrictive assisted reproduction technology law significantly improve pregnancy rate. *Human Reproduction, 26*(2), 376–381.

Slade, P., Raval, H., Buck, P., & Lieberman, B. (2007). A 3-year follow-up of emotional, marital and sexual functioning in couples who were infertile. *Journal of Reproductive and Infant Psychology, 10*(4), 233–243.

Stark, B. (2012). Transnational surrogacy and international human rights law. *ILSA Journal of International & Comparative Law, 18*(2). http://ssrn.corn/abstract=2118077

Surrogacy Arrangements Act. (1985). Legislation.Gov.UK. The National archives. http://www.legislation.gov.uk/ukpga/1985/49

Tallandini, M. A., Zanchettin, L., Gronchi, G., & Morsan, V. (2016). Parental disclosure of assisted reproductive technology (ART) conception to their children: A systematic and meta-analytic review. *Human Reproduction, 31*(6), 1275–1287.

The United Nations Convention on the Rights of the Child. (1989). http://www.unicef.org.uk/Documents/Publication-dfs/UNCRC_PRESS200910web.pdf. Accessed 19 Oct 2016.

The Universal Declaration of Human Rights (UDHR, 1948) Article 16.1. http://www.claiminghumanrights.org/udhr_article_16.html. Accessed 19 Oct 2016.

The US Fertility Clinic Success Rate and Certification Act. (1992). (FCSRCA) Pub. L. no −102 493.

UN General Assembly. (1948, December 10). Universal declaration of human rights. 217A (III). http://www.refworld.org/docid/3ae6b3712c.html. Accessed 9 May 2016.

UNESCO. (2006). Universal declaration on bioethics and human rights. http://unesdoc.unesco.org/images/0014/001461/146180E.pdf. Accessed 23 May 2016.

van den Akker, O. B. A. (1993). Prophylactic benefits of antenatal screening: Helpful or harmful? *British Journal of Midwifery, 1*(5), 220–223.

van den Akker, O. B. A. (2003). Genetic and gestational surrogate mothers' experience of surrogacy. *Journal of Reproductive and Infant Psychology, 21*(2), 145–161.

van den Akker, O. (2012). Chapter 10: Overcoming involuntary childlessness and assisted conception. In *Reproductive health psychology* (pp. 162–165). Chichester: Wiley.

van den Akker, O. (2016). Reproductive health matters. *The Psychologist, 29*(1), 2–5.

van den Akker, O. B. A., Crawshaw, M. C., Blyth, E. D., & Frith, L. J. (2015). Expectations and experiences of gamete donors and donor-conceived adults searching for genetic relatives using DNA linking through a voluntary register. *Human Reproduction, 30*(1), 111–121.

van den Akker, O., Postavaru, G., & Purewal, S. (2016). A systematic review and meta-analysis of the psychosocial consequences of twins and multiple births following medically assisted reproduction. *Reproductive Biomedicine Online, 33*(1), 1–14.

Van Zyl, L., & van Niekerk, A. (2000). Interpretations, perspectives and intentions in surrogate motherhood. *Journal of Medical Ethics, 26*, 404–409.

Wright, J., Duchesne, C., Sabourin, S., et al. (1991). Psychosocial distress and infertility: Men and women respond differently in vitro fertilization. *Fertility & Sterility, 55*, 100–108.

Young, I. (1990). *Justice and the politics of difference*. Princeton: Princeton University Press.

10

The Legal Framework

The various British committees and enquiries which have contributed to and drawn upon existing legislation are delineated in the final chapter. The costs and benefits of the role the legal system should play in surrogate motherhood arrangements are explored, including the possibility of opening surrogacy to criminalisation if laws are broken, and the severe and unintended consequences this can have on all parties involved. The consequences for a child born following a criminal or illegal act may be particularly serious as she/he will have to live with the criminal fact of their existence (see also the 1984 Warnock Reports' concerns about 'tainting the practice with criminality'). Criminalisation also breaches aspects of human liberty to procreate as individuals see fit. Finally, a proportion of surrogacy in the United Kingdom and elsewhere is carried out outside the medical arena via do-it-yourself surrogacy using artificial insemination which is legal but difficult to monitor—meaning it may not always be possible to determine what methods were used to conceive a child. Monitoring, accountability and accurate birth registrations are therefore of critical importance for the human rights to identity of the offspring.

Legal enforcement of surrogate motherhood arrangements also means that commissioning individuals may be forced to take on a baby they do not want and surrogate mothers may be forced to relinquish a baby they

© The Author(s) 2017
Olga B.A. van den Akker, *Surrogate Motherhood Families*,
DOI 10.1007/978-3-319-60453-4_10

do want. The psychological effects on the child of being unwanted, on a surrogate woman to be forced to relinquish her (genetic) child and on a commissioning individual or couple to be forced to take a baby they no longer want could be devastating. The effect on a genetic commissioning father may be equally devastating if the surrogate does not relinquish the baby. We have no understanding of any existing or potential emotional trauma in men providing gametes and their experience of regret or loss of an unknown child. Unfortunately, it is highly unlikely that research can determine who potentially suffers more loss—a genetic parent, a social parent or a gestational parent. Studies of gamete donors who have registered on a DNA register to find—and be found by—any offspring from their donations suggest some feel a responsibility to the child, and are happy to have some form of contact (Blyth et al. 2017).

Nevertheless, it is argued that some form of legal involvement is necessary in surrogate motherhood arrangements because this process is associated with multiple risks and involves several components, each related to each and every individual involved:

- The commissioning parents put in a great deal of emotional effort in trusting a surrogate to deliver to them what they could not deliver themselves: a baby. This emotional effort and investment should be protected from possible abuse. The most fearsome form of emotional damage to the commissioning couple would be through non-relinquishment of the baby by the surrogate mother.
- The commissioning parents will have to deal with the fact that the surrogate may be relinquishing a child to which she is genetically related which may become a traumatic prospect; the surrogate mother may also become medically, socially or psychologically compromised as a result of the surrogate arrangement, which could lead to feelings of guilt in the commissioning parties.

- The surrogate mother puts in an enormous amount of physical (medical) and socioemotional risk and effort to assist a couple receiving the baby.
- She will also have to deal with the fact that she will be relinquishing a child which is (in genetic surrogacy) genetically related to her,

to a couple she perceives as potential good parents at the start of the arrangement, but not necessarily at the end.

- A (gestational) surrogate should be protected from having to care for a baby she did not conceive for herself, whether it is or is not genetically related to her. This may be even more problematic if the child is found not to be entirely healthy.

- The surrogate baby should in law have a right to access regarding her/his genetic and/or gestational origins and has a right to a nationality.
- The surrogate-born individual may find out the surrogate mother was subject to multiple risks and was medically, socially or psychologically compromised.
- The surrogate baby, because of the 'deliberate' nature of its existence, should be protected from becoming an unwanted child, where neither party in the surrogate arrangement wants to take responsibility for it. This could be a more pressing issue if the child was not healthy or in cases of multiple births or the 'wrong' gender of the child(ren).
- The surrogate child should be protected from believing it was bought, and from believing it may not meet the purchase requirements the commissioning parents had hoped for.

These reasons are serious enough to warrant some form of effective but compassionate legal involvement. In accepting this premise, it may seem surprising to find that the laws and regulations in the United Kingdom and abroad have not taken to surrogacy in a systematic and committed way. The main reason for this slow-to-emerge legal commitment to surrogacy was probably because surrogacy was never envisaged as a large-scale issue in the United Kingdom. However, the incidence of surrogacy has increased. Now, many hundreds of people are turning to surrogacy as a means of overcoming childlessness, and they should not be left to their own devices for the reasons outlined above. It is generally recognised that some form of regulation is necessary. This recognition comes from the people involved in surrogacy (the commissioning and surrogate parents), the agencies involved in putting people together and the clinics which are involved in assisting in the surrogate conception. Moreover, professional bodies such as PROGAR (Project Group on Assisted Reproduction), BASW (British Association of Social Workers), the BMA and legislative bodies such as the HFEA (HFEA 1998) and the relevant government departments recognise this.

In the United Kingdom, surrogacy arrangements can break down and fail since the arrangement is not enforceable in law. A commissioning couple cannot apply for a PO unless the child is already living with them and unless the surrogate mother consents. Any court cases contesting ownership of the child are bound to favour the surrogate mother as birth mother, thereby enforcing the law of the land (rather than favouring genetic parenthood of the commissioning father or the mother). However, in practice this is not always the case (see Chap. 8). In cases of dispute of 'ownership' of a surrogate baby, motivations and emotions may have to be assessed on an individual basis. Even then the task of allocating a surrogate child may be difficult. Although the United Kingdom has passed surrogacy legislation which is probably the most comprehensive in the world and British law automatically confers parenthood status to a birth mother, in cases of dispute judges repeatedly allocate legal parenthood to parents who have not given birth to the baby (see Chap. 8).

Across the world, individual countries adhere to their own legislation, ranging from what are described as 'dangerously permissive' in some south-east Asian states (Ramskold and Posner 2013) to a complete ban on surrogacy as in, for example, Italy. However, individual legal systems often conflict on nationality and parenthood issues. Some form of international regulation is therefore necessary for cross-national surrogacy arrangements. A new internationally agreed regulatory framework would need to be agreed independent of the individual countries' existing frameworks, but with the ability to complement national variations in law. Some proposals also include an international certification scheme of fertility and surrogacy clinics (Donchin 2010) which will also assist in accurate record keeping. According to Ramskold and Posner (2013, p. 400) such a solution would gain in popularity if 'globally recognised accreditation was part of international law and guided by an amendment of the International Health Regulations International Health Federation (2005) of the World Health Organization (WHO) Constitution'. Ramskold and Posner (2013) further suggest that a separate convention under the HCCH—the WHO for cross-border Co-operation in Civil and Commercial Legal Matters—would help effect international guidelines (HCCH 2012) building upon existing inter-country agreements and inter-country adoption (HCCH 1993). Below is a summary of the major acts and committees which existed and informed, contributed or led to the current legislation on surrogate motherhood in the United Kingdom.

The Adoption Act (1976)

The 1976 Adoption Act was enforced to amend and extend the Adoption Acts 1952 through to 1974. Some of the requirements which must be met for an adoption to take place legally will be outlined in this section (for details of the full Act, refer to the original document listed in the reference section). The adoption Act allows people to apply for an adoption order. An adoption order allows the child listed on the order to be placed under parental responsibility to the applicant, and ends the previously responsible parent's responsibility for that child. Only children under the age of 18 years can be adopted, and any person who is single, a step-parent or a married couple over the age of 21 can adopt. If one of the married spouses is the parent of the child, she/he can adopt at the age of 18 years.

An application to adopt can be made only through an approved adoption agency, such as a local authority, although adoption by a relative of a child can be made through a High Court order. The primary consideration in any adoption is to safeguard and promote the welfare of an existing child and if the child is able, his/her wishes must be taken into consideration. The court will need to be satisfied that all parties involved in the consent to the adoption were fully aware and agree unconditionally to the proposed adoption. Critically, the birth mother must wait six weeks following delivery before she will be allowed to consent to the adoption. This safeguards her against making a decision she may regret later on, by allowing her time in the post-natal period to consider her decision. A number of other requirements must be fulfilled prior to adoption:

- Any child who is to be adopted must have lived with the adopting parent(s) for a period of time prior to making an adoption order.
- The adoption agency must have been in a position to observe the child with the applicants.
- The applicants have not previously applied for an adoption order, and been refused an adoption
- No unauthorised payments have been made.

There are some aspects to parental agreements in adoption which are also worth noting. For example, although it was previously stated that *all* parties involved must consent to the adoption, exceptions are made if:

- a parent cannot be found or is incapable of giving agreement;
- a parent withholds agreement on unreasonable grounds;
- a parent has persistently failed to discharge the parental responsibilities for the child (without reasonable cause);
- a parent has abandoned or neglected the child;
- a parent has persistently ill-treated the child;
- a parent has ill-treated the child and cannot rehabilitate the child within the existing household.

Once an adoption order has been placed, the child's *legal* relationship with the birth family is terminated, and brings to an end any previous court order in relation to the child. The Registrar General keeps the Adoption Contact Register, containing the name and address of any adopted person who has reached the age of 18 and wishes to have contact with a relative. A register for relatives of the adopted person also exists, so that the relative can contact the adopted person. The system therefore takes primary responsibility for the welfare of the child—including an open route to contact its genetic relatives—and considers the birth mother's potential to change her mind following the well-recognised hormonal tumultuous first six weeks following delivery. It also tries to ensure the new adoptive parents are suitable to adopt the child. Although a useful guide for surrogate-born children, clearly this Act is not designed to apply to the welfare of the surrogate-born child as she/he is not already existing but instead is specifically commissioned for a person or persons. Nevertheless, the six-week contemplation time for the birthmother, the observation of the adoptive family with the child and the open route to tracing and/or contact with genetic relatives are key to successful adoption procedures, a model befitting a surrogate-born child.

The Warnock Report (1984)

In 1982, a Committee of Inquiry into Human Fertilisation and Embryology was set up as a result of public concerns about the consequences of advances in new reproductive technologies. Their examination considered the social, religious/moral, ethical and legal implications of current and future developments in human reproduction including surrogacy and what policies and safeguards should be applied. The report from that committee resulted in the Warnock Report (1984), which was intended to provide a reasoned discussion of the issues, encourage a high-standard public debate and guide subsequent public policy. The Government implemented some of the recommendations made by the Warnock Committee into the Surrogacy Arrangements Act (1985) and some within the HFEA Act (1990). The context in which the 1985 Warnock Committee was holding its discussions was one where no provision for surrogacy was available on the NHS; only one agency had started to operate on a non-commercial basis in the United Kingdom (COTS); and this practice was not unlawful. The practice would have been deemed unlawful, only if according to the 1958 Adoption Act (Section 50) payments had been made in connection with the adoption.

There were strong arguments against surrogacy in the Report, including the following:

* Surrogacy could be seen as an aberration to the value of the marital relationship by the involvement of a third party in the procreative process normally limited to two people. Its third-party involvement would, unlike in gamete or embryo transfer, involve a third party in a more intimate way.

- The use of another woman's womb for financial gain is inconsistent with human dignity.
- A breach of the 'mother-child' relationship is introduced in surrogacy which could be damaging psychologically to the child, or indeed be considered degrading if it was to be considered as a commodity in a financial transaction.
- No woman should undertake the risk associated with any prospective pregnancy for another woman in order to earn money.

In essence, the arguments against surrogacy rested on the assumptions that to introduce a third party into the intimate process between two people within a marital relationship is an attack on the marital relationship. This attack is worse than the introduction of, for example, a semen donor carrying AIDS into the relationship, because a surrogate mother's involvement is more intimate and personal; to use a uterus for financial profit is against human dignity; a woman willing to conceive a child for the sole purpose of giving it up after delivery strengthens the human dignity argument and is seen as the wrong approach to pregnancy; the intent to give up a baby following delivery can be potentially damaging to the child as the child can perceive this as being bought for money; no woman should carry the risks of pregnancy for another for financial gain and no woman should be forced by legal means to give up the baby she delivered.

In support of surrogacy, the Warnock Report argued:

- Surrogacy could be seen as a deliberate and thoughtful act of generosity on the part of one woman to another.
- There is no evidence demonstrating any form of psychological damage to a child upon separation from the birth mother. Even if any damage was identified, the same is already accepted in adoption.
- Any woman has a right to enter a surrogate relationship if they desired, there are no grounds for believing anyone did this lightly, or that payment affected the voluntary intent of the agreement.

The arguments in favour of surrogacy included the fact that if this is the only 'life enhancing' option for infertile couples with no other options, in particular now that a full genetic link was possible through IVF surrogacy, surrogacy should not be ruled out; surrogacy can be seen as an altruistic act; women have every right to enter into such an agreement; if the agreement is entirely voluntary, there is no reason to assume exploitation or women entering these agreements lightly; as no evidence exists what, if any bonding, takes place between the unborn baby and the surrogate, no great claims can be made that breaking these bonds is harmful.

At the time this committee was set up, Britain had few known cases of surrogacy to draw on. The information regarding surrogacy came largely from abroad, influencing the recommendations made at the time, but not reflecting surrogacy practices in the United Kingdom. Some strong statements were made; for example, surrogacy for convenience is 'totally ethically unacceptable'; the danger of exploitation, even if surrogacy is commissioned on medical grounds, appeared to outweigh the potential benefits to the commissioning couple 'in almost every case'; surrogacy was seen by many as 'positively exploitative when financial interests are involved' (paragraphs 8.17 and 8.10–8.12). Moreover, the commissioning father, almost always married to the commissioning mother, may be subjected to an affiliation claim by a surrogate mother if she keeps the child. He will be regarded in law as the unmarried father of the child and therefore be subject to all the conditions other unmarried fathers have to face up to.

Following the debates, the Warnock Committee concluded that:

1. Legislation be enacted making it a criminal offence to create or operate in the United Kingdom any agency whose purposes include 'the recruitment of women for surrogate pregnancy or making arrangements for individuals or couples who wish to utilise the services of a carrying mother' (paragraph 8.18).
2. Another interesting caveat in the conclusions was that legislation should make it clear beyond any possible doubt that surrogacy agreements are illegal contracts and therefore unenforceable in courts. However, criminal law should not be used to stop individuals entering into private surrogate arrangements, to 'avoid children being born to mothers subject to the taint of criminality'.
3. No express judgements on payment were made as this would not be acceptable.
4. A non-profit-making surrogacy service would 'in itself encourage the growth of surrogacy'.

Other proposals made by the Warnock Committee, based on a minority of voices, recommended that:

1. A statutory authority should be created to regulate infertility treatment, including surrogacy.
2. The authority would have the power to license non-profit-making agencies who wished to assist in making surrogacy arrangements.
3. Access to a surrogacy agency would be only by referral from a gynaecologist.

This 'minority' view from the Warnock Committee concluded wisely that a final judgement on the ethical and social dimensions of surrogacy was not within their power. Rather than judging or suggesting a legal ban on surrogacy arrangements, they suggested that 'the door be left slightly ajar so that surrogacy can be more effectively assessed' (p. 89). In summary, the Warnock Committee was not unanimous in their conclusions or recommendations. It is likely that the lack of evidence demonstrating harm was responsible for the equivocal nature of the report.

The Surrogacy Arrangements Act (1985)

This Act was established following the Warnock Committee Report and following the first public outrage at surrogacy in the United Kingdom following Kim Cotton's relinquishment of 'baby Cotton' (see Chaps. 1 and 4). It outlawed commercial surrogacy. From a historical perspective, around the mid-1980s, some official recognition of the occurrence of surrogacy in the United Kingdom was made public, although it was a peripheral involvement. In 1985, the Surrogacy Arrangements Act simply stated that:

• Commercial surrogacy arrangements are prohibited by law
• Advertising for surrogates is prohibited by law
• Non-commercial surrogate agencies or individuals working to establish surrogate arrangements are not prohibited
• Payment to a surrogate is not prohibited

No other issues were explored, and no regulation was deemed necessary. It was assumed that if individuals wished to pursue surrogacy they would not be prevented from doing so, provided they did not do it on a commercial basis. The 1976 Adoption Act was in place and could deal with surrogacy cases in a way similar to adoption. Thus, if both parties agreed to the legal adoption of the child, this would be possible through the already existing Adoption Act. What was not foreseen at the time were the many serious issues arising from adopting within surrogate arrangements (Andrews and Douglas 1991). In turning to the 1976 Adoption Act, the parties involved in surrogacy were faced with unnecessary complications.

- A surrogate baby would automatically be registered as the legal child of the surrogate couple. Both the surrogate and *her* husband (who is never likely to have had any input into the surrogate conception) could be registered as the baby's legal birth parents.
- A Parent Responsibility Agreement can be entered into by the commissioning father and the surrogate mother once the child has been born. This agreement gives these two people equal rights over the child, until the commissioning parents apply for a PO to give them full and permanent rights over the child.
- Six weeks after the baby is born, the commissioning parents would have to apply for a PO to reregister the baby in their name. This is the case, even if the baby is entirely genetically made up of the commissioning couple's egg and sperm.
- A surrogate birth mother will *always* be recognised as the legal parent of the baby, following delivery of the baby.

The Glover Report (1989)

Jonathan Glover, a moral philosopher, was invited to chair a European commission to examine the ethical, social and medical problems generated by the new reproductive technologies. The reason was because although national discussions had already taken place in some European countries, a shared European policy could be appropriate with the (at the time) gradual unification of European Community countries. As Glover notes 'it may seem increasingly absurd if, say, surrogate motherhood is banned in one place but allowed half an hour's flight away' (p. 15). The Glover committee's brief was relatively open, so they concentrated their efforts on four main points: (1) the three main groups involved—parents, donors and children; (2) surrogate motherhood; (3) embryo research (4) and the implications of how the new reproductive techniques can influence the kinds of people who are born. This section will outline some of the main discussions held and conclusions reached about surrogate motherhood (for details of the remaining three topics discussed by the committee, see Glover 1989).

The Glover committee made excellent use of the then available evidence to inform them of current (1980s) practice and research. Apart from the obvious issues which could serve conflicting interests such as exploitation for financial reasons, they noted a few issues apparent during a surrogacy arrangement which are contrary to a surrogate's human rights. These included the contractual (albeit unenforceable) stipulations made by commissioning couples:

- Sex during the insemination period with her own partner(s) is not allowed.
- A blood test upon the baby's birth will be carried out to check that the child was conceived from the commissioning husband's sperm.
- Medical examinations, including amniocentesis, may be required during the pregnancy.
- The surrogate may be expected to have an abortion if the ante-natal test for foetal abnormality is positive.
- Because the child is 'explicitly defined as not hers' (p. 70), she may not dispose of herself or the child during the pregnancy as she sees fit.

The surrogate mother is therefore not autonomous as she is not in a position to make decisions for herself, about herself and the unborn child she may be carrying. This leads on to the two main ethical issues discussed in this report. Firstly, is surrogate motherhood morally acceptable and secondly, should it be legally permitted? Their case *for* surrogacy can be summarised as: it relieves the burden of childlessness. Against surrogacy, a number of valid concerns were listed:

* Conflicts arising between the surrogate mother and the commissioning parents. This conflict could harm the child.
* Introducing a third party into the traditional intimacy of conception may weaken the institution of the family.
* Financial pressures may weaken a surrogate mother's position to resist contractual conditions which give little weight to her interests.
* It is an invasion of a surrogate's bodily integrity.
* Subsequent regret at having given up the baby.
* Ill effects upon the surrogate's own children.
* A surrogate may misleadingly believe she is developing a special friendship, which is dismissed by the commissioning couple following completion of the arrangement.

The resultant Glover report endorsed a restrictive approach to surrogacy. Some of their conclusions are summarised below:

* It was important to protect surrogate mothers from exploitation.
* It stated that children should be protected from harmfully prolonged battles between the surrogate and commissioning parents.
* Commercial agencies should be prohibited.
* Surrogate mothers should not be bound contractually to relinquish the child.
* Payment beyond expenses should not be encouraged.

Interestingly, it was not within their conclusions to state that when surrogacy is an only option for a commissioning couple to have a child, public agencies could assist in surrogacy arrangements. The Glover report also did not proclaim that payments should be illegal. Their main reason for deciding that contracts should be unenforceable against a surrogate mother, apart from protecting her, was to ensure that a surrogate, if she changed her mind, would not become depressed and therefore not have an adverse effect upon the unborn child. Another point made was that without denying the possible importance of a genetic link of the commissioning father with the child, his genetic link should not trump the other biological bonds (of pregnancy and delivery in gestational surrogacy and pregnancy, delivery and a genetic link in genetic surrogacy). It was decided that it would be better for the commissioning couple to accept the fact that a surrogate could change her mind about relinquishment, than to force a surrogate to endure the anguish of being made to give up the child she gave birth to. From the child's point of view, making a contract enforceable in law would make breaking it a criminal offence. This is undoubtedly not good for a child and neither are prolonged legal battles.

Some screening was recommended by the committee, for both the surrogate mother and the intended commissioning couple. They dealt with this issue as follows. If agencies or clinics were made responsible for implementing some form of regulation, they could operate on guidelines and a licence and be publicly inspected. Screening of both parties involved could benefit all concerned, and this could be part of the brief of the licensed agencies or clinics. Screening is not well defined, but is used intermittently with counselling. Any counselling or screening would benefit the surrogate mother by giving her time to reconsider surrogacy, and to protect any potential child from being carried and nurtured by heavy smokers, alcoholics or drug addicts. Interestingly and of substantial concern is the statement that they accept there may be grounds for overriding these reasons for exclusion as part of the screening recommendations if, for example, surrogates are hard to come by. They justify that by making the risky statement that from the perspective of the child, it may be better to have run those risks, or even to have suffered some harm, than not to have been born at all. Future claims against epigenetic harm caused by (surrogate) mothers may show this justification was not in the best interests of the child, or of the parents caring for a child with serious disabilities (such as foetal alcohol syndrome) as a direct result of the poor in vivo environment provided by the surrogate mother. The evidence for the long-term effects of the gestational environment is increasingly recognised (Gardner and Lane 2004; EpiHealth 2016).

Another surprising recommendation made in the report concerns the suggestion that severance of the surrogate mother relationship is regarded as a good thing following relinquishment of the baby. The committee pointed out that visits or other contact between the surrogate mother and the surrogate child could undermine the child's security about who his or her parents are (although an exception is made within arrangements made between relatives). This was not evidence based and current practices (evidenced by COTS annual meetings) and research on adoption and donor offspring report the opposite—as shown below. They also suggest that 'there is a case for the child having the right to be told on reaching maturity the identity of the surrogate mother.' Disclosure of origins is necessary and was a positive recommendation from the report. However, leaving disclosure until the person reaches maturity goes against all good practice and evidence showing that if individuals know from their earliest years that their conception was different (as is also demonstrated in advice given in adoption and more recently donor conception, Freeman 2015), the effects of the disclosure do not harm family relationships (Freeman 2015) as was previously thought. A commonly reported response among those who are told of their non-traditional conception at a young age is curiosity (Jadva et al. 2009) and having a neutral to positive impact on parent-child relationships (Blake et al. 2010; Mac Dougall et al. 2007).

The committee also stressed that payment for donation of any kind, be it semen or blood donation or surrogacy, affects the *psychology* of donation, 'turning what could be an enriching act of altruism into an act more like selling an old motorbike' (p. 84). Although the committee strongly favoured a non-commercial basis for donation, they did not believe a legal ban to commercialism was the answer either. The Glover report emphasised the fact that in surrogate motherhood much more than donation is involved, including a great deal of inconvenience and 'some' risk. Their argument is one where payment may lead to exploitation of poorer women. Substantial payment may, on the one hand, make the arrangement less exploitative, but at the same time may make it harder for a woman needing money to resist. However, considering higher pay may be less exploitative and less pay more exploitative, if the higher pay puts more pressure on a surrogate to do it for the money which she may later regret less pay seems the lesser evil.

Thus rather than recommending a commercial approach, an option with no or minimum pay was seen as preferable even if that means less surrogates will offer themselves for surrogacy. The committee was therefore more concerned with payment than with quality of the surrogate mothers in vivo environment, and was willing to compromise on the child's health but not on payment if it meant less surrogates coming forward. This implies the welfare of the commissioning parent(s) was placed before that of the surrogate mother and before the welfare of the child. In the case of agencies, the committee based their argument for non-commercialism largely upon the argument that this would minimise 'the encroachments of commerce on the intimate relationships of parenthood'. The main reason for approaching the issue on this basis was because commercialism changes the way the surrogacy is perceived, as buying and selling transactions. The conclusion of the committee regarding agencies thus rests on the premise that 'because the commercialisation of intimate relationships seems something to resist, and because restricting liberty of commercial organisations seems less intrusive than restricting the liberty of individuals, we do not favour permitting commercial agencies for surrogacy'. In sum, the Glover committee produced a well-argued, thoughtful report containing many worthy recommendations. Despite the quality of the report, the European view generally remained one of restrictions on surrogacy (France, Denmark and the Netherlands), or altogether a ban on surrogacy (as is the case in, e.g. Austria, Germany, Sweden and Italy). The European community therefore did not share the recommendations of the Glover report.

The British Medical Association (1990, 1996)

In 1990, the BMA published a report on surrogacy (Surrogacy: Ethical Considerations 1990) reporting a fair degree of resistance of the medical professionals' involvement in IVF surrogacy. The view expressed was not entirely favourable towards surrogate motherhood practices, although they recommended that some degree of professional assistance should be provided. In 1996, the BMA published an update to the 1990 report, entitled 'Changing conceptions of motherhood: The practice of surrogacy in Britain'. This publication acknowledges the different attitudes held in their previous report regarding surrogacy.

Instead, they urge for more research to be carried out, safeguarding accurate evidence-based documentation on surrogacy as it takes place in the United Kingdom today. They also announced that surrogacy in their view is now regarded as 'an acceptable option of last resort'. The main contribution made by the BMA as a professional body was that they pointed out there was a total research vacuum with regard to the surrogate mothers', the baby and the commissioning parents' mental and physical health and wellbeing.

Guardians ad Litem (Local Authority Circular) LAC (94)25

Prior to the 2008 Act, and the Parental Order Reporters (PORs; see section below), guardians were appointed to ensure the welfare of the child subject to a PO is first and foremost ensured. They were also required to check that no payments other than expenses were made to the surrogate. A guardian ad litem is an independent social work professional appointed by the court to represent a child subject to care or adoption proceedings. They are not already involved in the case they are appointed to and do not work for the local authority.

The function of the guardian was to provide the court with an independent view of what has been happening and what should happen in a child's life. These court appointments were usually made under the Children Act 1989, hence the term 'guardian ad litem' actually means guardian of the proceedings. A guardian ad litem normally learns what she/he can about a child, their needs and their care, and visits tend to be made to obtain an independent view of the child's situation and to develop a file on the case for the court. A guardian works closely with the child's solicitor, and both the guardian ad litem and any other witnesses may have to give evidence on the child's behalf in court. The report compiled by the guardian remains confidential, and belongs to the court, although copies are sent to the child's solicitor, the parents and the local authority. This also implies that the guardian ad litem can be called as witness by the parents if they disagree with the evidence contained in the report.

What was a peculiar outcome of the law (or lack of law) is that by the time a guardian has been appointed, the child is already living with the commissioning couple. Considering their main brief was to consider the child's welfare, it could be argued that their appointment came a stage too late. It was often the case that guardians ended up in retrospect assessing the implementation of Section 30, limiting their powers considerably, not least because it would not be in the child's best interests to move her/him away from its intended parents after bonding has likely taken place. Unlike in adoption, no police checks are carried out on commissioning couples and no checks are made available to ensure at least one of the intended parents is in fact genetically related to the child. Thus, a guardian's ability to carry out the necessary scrutiny of a couple's suitability is severely hampered.

Parental Orders (2010)

As previously noted, in the United Kingdom the birth (surrogate) mother is registered as the legal mother of the baby (see, e.g. 2008 Act, section 33). A commissioning woman under this same Act cannot be treated as the mother even if her own ova are used in the conception of the surrogate baby via gestational surrogacy (2008 Act, section 47). On the other hand, a commissioning father is recognised as the legal father of the baby but if he did not use his own sperm, the surrogate's husband—if she has one—will be registered as the legal father of the baby—not the sperm donor. In all cases where parenthood is legally attributed to the surrogate or the surrogate and her husband, the commissioning parent(s) need to apply for a PO to acquire the legal status as the parent of the surrogate baby. The Family Proceedings Court appoints an officer from the Children and Family Court Advisory and Support Service (Cafcass) to act as a POR and assess if the child's best interests are served by an order. PORs are qualified social workers and they investigate the financial agreement between the commissioning parents and surrogate mothers and prepare reports for court (Cafcass 2010). According to the Department of Health, a decade ago 50–70 PO applications were made yearly (Department of Health 2010). It is of substantial concern that there is good evidence to show not all parents of national and international surrogate-born babies register for a PO (Crawshaw et al. 2012a).

In a rare study of PORs attitudes towards and experiences of being a POR in surrogate cases, Purewal et al. (2012) reported largely positive attitudes towards surrogacy. PORs expressed some certainty that openness about the surrogate origins was in the child's best interests but there were concerns about parental preparation and assessment arrangements; particularly overseas arrangements and concerns over the non-regulation of surrogacy agencies were also expressed. Of concern also were the high rates of role ambiguity, as this led to reports of less positive attitudes towards the emotional consequence of surrogacy on offspring. Furthermore high scores on role conflict and role ambiguity were also linked to less positive attitudes towards the parties' preparation towards parenthood. Clearly training, policy and practice would benefit from inter-disciplinary and inter-agency linking, monitoring and record keeping.

Hodson and Bewley (2017, p. 1) pointed out that there are some substantial problems with POs, relating to breaches in human rights and ignorance relating to POs even amongst professionals facilitating surrogate pregnancies. For example, although same-sex couples now can apply, single men still cannot apply for POs which constitutes an 'infringement of the prohibition of discrimination with the right to a family life'. Furthermore, a surrogate mother can consent to commissioning couples raising a surrogate-born baby but not consent to the PO, leaving the law in a peculiar state of ineffectiveness. This may be important if she is not certain that the right decisions will be made by the commissioning parents about the best interests of the child. However, replacing the imperfect PO system with a new system of pre-birth orders could result in a practice where babies are forcefully removed from the surrogates, increasing the stigma associated with the practice and decreasing the rights of the surrogate mothers which could be 'perpetuating violence against women' (Hodson and Bewley 2017, p. 2). It is imperative that discrimination is removed from the current system, and that no further derogatory practices for the surrogate mother take its place. Pre-birth orders go against the evidence and years of good practice eventually put into legislation in adoption. It is inconceivable that this is ignored in future changes in legislation of surrogacy.

The Human Fertilisation and Embryology Act (1990, 1998, 2008)

The HFEA is the regulatory body for good practice of ART in the United Kingdom. It submits reports (and regulations) to all organisations involved with assisted conception and provides information to the population on what is currently considered good practice. The HFEA (HFEA 1990) was given the responsibility of regulating some fertility treatments. These included regulating the use of donated sperm, eggs or the creation of embryos in vitro. If treatment centres used any of these they had to apply for a licence from the HFEA, and treatment was to be carried out by a responsible person. Surrogate conceptions in clinics would therefore come under the HFEA's regulations, since insemination (as in genetic surrogacy) and embryo transfer (as in gestational surrogacy) are part of the licensed treatment offered. Consequently, the HFEA also indirectly regulates surrogacy if the surrogate conception takes place under the care of these licensed clinics. This conflicts with attempts to refer to surrogate arrangements as social contracts, as medical involvement should be recognised as partly responsible for the bringing about of the pregnancy and baby (van den Akker 2012, 2013, 2016).

However, all genetic surrogacy taking place outside of clinics (which was the case in the majority of genetic surrogacy cases) are therefore not regulated. The lack of regulation in do-it-yourself surrogacy arrangements has some advantages. Firstly, it is far cheaper to carry out the simple procedure of artificial insemination in the home. Second, it has less of a medical event feel to it, therefore being more akin to 'natural' conception. However, there are also some disadvantages associated with unregulated surrogate conceptions. Some of these can have serious consequences for all parties concerned, for example:

- A surrogate could become infected through diseased sperm from the commissioning husband.
- A commissioning couple can never be sure that the commissioning husband's sperm was in fact responsible for the pregnancy (unless they opt for genetic testing for confirmation of genetic fatherhood).

- A surrogate baby could sue either the commissioning or surrogate parents, once it reaches the age of maturity (18) for carrying a condition which was not tested for and/or disclosed at the time of conception.

These are not negligible concerns and could be avoided if insemination had taken place under a clinic licensed by the HFEA. As their procedures are regulated to ensure all care is taken to transfer only quarantined and tested sperm, detailed histories are taken revealing any conditions which may result in genetic transmission of a condition, and the clinic is responsible for inseminating the commissioning husband's sperm. Thus, safeguards are far more substantial in licensed initiation of genetic surrogate conceptions than in do-it-yourself genetic surrogate conceptions; the latter are effectively entirely 'social' contracts.

From a psychological and social point of view, clinics are also required by law to safeguard the wellbeing of all parties concerned. The primary concern focuses on the wellbeing of the child. In practice, this means that if a clinic perceives a couple to make good parents, then they have considered the wellbeing of the child. Criteria for meeting the definition of 'good parents' are entirely at the discretion of the professional(s) involved, although they tend to include aspects of health and age of the prospective parents. In theory good parenting is far more complicated than this, but clinicians are often unwilling to deny hopeful prospective parents their desire to have a child. Unfortunately, there are couples who renege on their agreements. It is possible that some of these problems could have been avoided if adequate psychological screening had taken place. Currently there is no provision for this.

The HFEA does state, however, that a licence will only be provided to centres which provide the opportunity for counselling to be made available. This Act has left some gaps open to misuse. For example, a clinic would be following the Act to the letter by providing the opportunity for counselling, and no one would be breaking the code by not taking this opportunity up. Moreover, the Act (1990) S 13 (6); Schedule 3 Paragraph 3(1)(A) also states that the counselling provision should consist of 'proper' counselling availability to those born from assisted conception services (i.e. the offspring) who wish to obtain information about their genetic origins. This particular clause did not take effect until 2008, and nothing is known about who will be made responsible to inform the offspring, provide such counselling or who monitors the effectiveness (Crawshaw et al. 2012a, 2012b). Another problem arises concerning interpretation of the Act. Proper counselling is not defined, although guidance about the qualifications and experience of counselling staff and the nature of counselling is provided (HFEA 1998, 1.10).

Regarding the surrogate and prospective commissioning parents involved, the counselling required can be of three types: supportive, implications and therapeutic counselling. However, in practice, a counselling session is often deemed necessary by the clinical staff only for problematic cases (Blyth and Hunt 1994). This is not at all what counselling is about and it also blurs the boundaries between counselling for the benefit of the individuals concerned, versus some sort of assessment by a counsellor for the benefit of the clinical team's doubts about a person's suitability to receive the treatment. The current position is therefore that surrogacy is legal in the United Kingdom provided no payments for fees are made to a surrogate mother. Reasonable expenses are allowed, and although there is no formula for 'reasonable expenses' any costs incurred by a surrogate can be interpreted as constituting reasonable expenses. It is in fact left entirely to the discretion of the surrogate and commissioning couples involved in a surrogate agreement, although surrogate agencies such as COTS and Surrogacy UK provide some guidelines as to amounts which seem reasonable.

Other legally relevant points worth noting include the following:

- A commissioning parent was not entitled to maternity leave until recently, when this was changed to entitlement to parental leave or pay for both parents.
- No one is allowed to advertise for surrogates. Clinics and agencies have to take approaches from them, rather than encourage approaches formally.

Some commissioning mothers breastfeed their babies through appropriate hormonal stimulation and, like the surrogates who gave birth to the babies, are now given maternity leave or pay, just like any other person becoming a parent (see, e.g. Bowcott 2013). The decision by the European Court of Justice was based upon the case of a British commissioning mother who had started to breastfeed the surrogate-born baby. The European Court of Justice encouraged the UK Government to consider the reforms to UK maternity leave, which it successfully reformed (Maternity Action 2016). Employers will need to amend their family leave policies accordingly. However, this also means parents will have to disclose surrogacy in the workplace, a step many may be reluctant to take as was found in a study of disclosure of IVF in the workplace (van den Akker et al. (accepted for publication); Payne and van den Akker 2016). The lack of recruitment options may prove limiting in the end, and not necessarily assist in getting the best surrogates to come forward.

Section 30

This section was not inserted into the HFEA Act until 1994 and was added later because practice problems became evident (e.g. a commissioning couple disputed the fact they had to 'adopt' their own child). The couple, who had undergone IVF surrogacy using their own gametes, believed it was not acceptable that this child which was genetically theirs should be adopted by them. The Government supported their proposal and issued an amendment to the HFEA 1990 Act, including:

- All commissioning couples applying for a PO must be over 18 years of age, married and at least one of the couple had to be resident in the United Kingdom.
- At least one of the couple had to be genetically related to the child.
- The child must already be living with the intended couple and consent must have been obtained from the surrogate mother and the child's legal father.
- The court must be satisfied that 'no money or other benefit (other than expenses reasonably incurred)' has been paid to the surrogate, unless authorised by the court.

This changed again later to (PO Regulations 2010 and the HFEA Act 2008, section 54) a number of conditions which had to be met for legal parenthood to be assigned:

- The commissioning couple must be married, in a civil partnership or cohabiting and both parties must be >18 years old.
- The baby is genetically related to at least one of the commissioning parents.
- The PO application must be made within six months after the delivery of the surrogate baby.
- The baby must be living with the commissioning couple; one or both of them must be domiciled in the United Kingdom.
- The surrogate (and legal father if this is not the commissioning father) must give consent to hand over the baby within six weeks after the birth of the surrogate baby.
- No payment should have been made to the surrogate mother other than reasonable expenses accepted by the court.

The fact that same-sex couples can also apply for POs reflects the changing values in society about what constitutes the welfare of the child. This no longer includes the previously supposed need for a father and a mother; instead, a requirement for 'supportive' parenting is now in situ (Norton et al. 2013). Although British legislation is detailed and specific, there are some anomalies which require further consideration:

- An IVF couple using donated embryos are considered the legal parents.
  - In surrogacy at least one genetic link with the commissioning parents is required for them to qualify for legal parentage.
- Only couples can apply for POs, not single or non-genetically related parents.
  - Single parents can adopt, and be the legal parent of a donor insemination baby.
- Single parents can legally use surrogacy, but they cannot legally become the single parent of a surrogate baby.
  - The surrogate baby of a single parent, unlike her couple counterpart, cannot benefit from proper recognition that the single parent is the legal parent.

- A commissioning parent cannot make medical or legal decisions about the baby until the PO is granted, which can be up to six months after delivery.
  - The commissioning couple has the surrogate baby living with them in most cases since the delivery, so from a nurturing perspective, they are the parents but legally they do not yet have full parental responsibility.
- In international surrogacy cases the same British legal principles apply.

  - These British laws are discordant with most other countries which adhere to their own laws (see chapter on international surrogacy, Chap. 8).

Some countries operate using pre-birth contracts, bestowing legal parenthood to the commissioning couple as soon as the surrogate conceives (e.g. the Ukraine, Family Code of Ukraine Clause 123, 2008) or between the fourth and seventh months of pregnancy (e.g. as in California, Californian Family Code 2013). The problem with these arrangements is that they give the surrogate no opportunity to make a fully informed decision about her willingness and ability to relinquish the baby upon birth, a certainty she should be able to have, just as was recognised in adoption.

In 2013, the HFEA Code of Practice was introduced to assist fertility clinics, health care professionals, surrogates and intended parents with a series of guidelines:

- The surrogate mother is the legal mother of the baby upon delivery of the surrogate baby.
- If the surrogate is married, she and her husband or civil partners are the legal parents of the surrogate baby.
- When a surrogate mother is unmarried, or has a partner who does not consent, there are two options for second parent—the commissioning father is the legal father if no HFEA forms are signed indicating the contrary:
  - The commissioning mother's name can go on the baby's birth certificate. Both the surrogate mother and the commissioning mother need to sign the HFEA Parenthood Election Forms before conception.

- The commissioning parent can nominate a man who is not the biological father (e.g. if donor sperm is used or if the couple is same sex).
- If no parental election forms are signed and the commissioning father has provided the sperm, he can be named on the birth certificate as the legal father.

The legal position is important for numerous reasons, and must accurately reflect the coming about and legal responsibilities for the surrogate baby. First, the baby has a basic human right to grow up knowing its genetic origins and gestational origins. This is becoming increasingly more important in a society where genetic knowledge for health reasons is a human right (Harper et al. 2016), and the gestational environment can contribute in epigenetic terms to the expression and/or development of future conditions within the surrogate-born person (Ombelet et al. 2005; Egliston et al. 2007). For example, as noted in Part II, the surrogate mother influences the foetus via microchimerism (transfer of cells via the placenta between the foetus and the gestational mother), and trans-placental movement of antibodies and nutrients are all increasingly recognised as contributing to the developing foetus's own composition. Notwithstanding this, the post-natal environment usually contributed immediately after birth by the commissioning parents also has epigenetic influences upon the baby. Accurate legal recognition of the multiple parents involved in surrogate arrangements is also important from the perspective of succession rights, pensions and taxation.

British Infertility Counselling Association (BICA)

A number of articles by members of BICA have appeared, specifically relating to the regulatory aspects of counselling in infertility. Chartered infertility counsellors take part in official HFEA inspections and offer counselling to all parties involved in assisted conception particularly when using third parties. This shows the HFEA is taking the psychological welfare of the clinical patients undergoing fertility treatment in surrogacy very seriously. However, uptake is not mandatory and funding is generally not provided for counselling, leaving those in need with difficult decisions to make about allocation of resources. When access to funded treatment and associated supports such as counselling is limited, the financial impacts of treatment may become problematic for many people. This may be in part a contributing factor to the generally low uptake of counselling across all infertility treatment centres (Payne and van den Akker 2016). BICA (2016) reports on the guidelines for HFEA inspectors relating to infertility counselling, including:

- Is counselling routinely offered in accordance with the HFEA's code of practice in all clinics?
- What is the theoretical orientation of the counselling service?
- What safeguards are there to ensure counselling is not performed under the disguise of information giving and medical consultation?
- Does the centre offer support and/or therapeutic counselling in addition to the statutory requirements to offer implications counselling, and who provides this?
- Do patients receive both written and verbal information upon referral and at what point is such information given? How is the counsellor contacted by the patient?
- What level of counselling provision is made, what is the policy about routine referrals, ongoing counselling and what are the numbers of counselling hours in relation to numbers of patients treated?
- Is the accommodation for counselling suitable, is it a quiet, comfortable space free from interruption?
- How is the counselling service audited, what are counsellor uptake rates? And what other information is available on how the service is evaluated?
- What are the boundaries of confidentiality between the counsellor's work and that of the treatment team? Including information on written records of counselling—where are these kept and who has access to them?
- Has the counsellor any other additional roles within the organisation and is this appropriate or is there the potential for role conflict?
- How is the independence of the counselling service safeguarded?

- What is the centre's policy on the welfare of the child(ren) that may be born and any existing children and the role of counselling staff in the assessment of a patient's suitability for treatment?
- What are the centre's communication and decision-making procedures including the regularity of team meetings and whether counselling staff attend?
- If counselling staff are not members of the clinic team or are not working in the centre, what is the nature of their liaison with the team, their understanding of issues related to infertility and its treatment and how they remain informed of new developments?
- Are supervision arrangements adequate for counsellors, and is this cost covered by the centre?
- What are the procedures for referring patients to other specialist counselling or support services?
- What is the centre's policy on continuing education for counselling staff including information on courses undertaken and how are these costs met?
- Information about patient support groups and their relationship with the counselling service.
- What availability is there for patients about local/national support and counselling services?

Counselling is an underused resource in assisted conception as evidenced in a recent Fertility Network UK (FNUK) commissioned report of 780 members of FNUK, National Gamete Donation Trust (NGDT), Male Fertility UK, Care Fertility, COTS, Surrogacy UK and The Baby Centre (Payne and van den Akker 2016). In the survey, 44% of those responding received counselling; 54% of these had to fund some of it themselves. However, counselling was clearly seen as an important resource since 75% would have liked to have had counselling if it was free. It was not only formal counselling which was deemed to be necessary. Only 17% of respondents attended a support group, yet 52% would have liked to attend had there been one nearby. Since counselling also offers informational guidance, opportunities to discuss the implications of the treatment and to more fully understand the treatments are also lost due to lack of availability and funding; for example, only 52% of respondents fully understood the nature of their fertility problem and only 26% felt their GP provided sufficient information. Critically, mood was low in those completing the FNUK survey, with sadness, frustration and worries reported frequently and a staggering 42% experienced suicidal feelings as a result of infertility and/or treatment. Similarly, 70% reported some detrimental impact of infertility and/or treatment on their relationship, and 50% felt concerned that treatment would affect their career prospects, or that their career was damaged (33%) as a result of the treatment.

The Brazier Report (1998)

In 1997 a prominent group of people were invited by the then Minister for Health, to review surrogate arrangement practices in the United Kingdom. The team presented their findings in what is referred to as 'The Brazier Report' (1998). The brief was to determine if the law continued to meet public concerns. The focus was on payment, the need for a regulatory body, and if changes were needed to the Surrogacy Arrangements Act 1985 or the HFEA 1990. The results were various and are summarised below:

- They noted that in their view the recommendations of the Warnock Committee were incomplete.
- Documentation should be available of the number, type and outcome of surrogate arrangements in the United Kingdom, and of the welfare of the children born to surrogate mothers.
- Regulation of surrogacy is recommended.
- Payment to surrogates should be no more than genuine expenses, should be statutorily defined by a new surrogacy act and should not involve payment for 'services'.
- Agencies should operate on a non-profit basis and should have the expertise to help in setting up any arrangements and monitor the activities of the agencies.
- The UK Department for Health, in consultation with the HFEA, should develop a Code of Practice to set out minimum standards to

minimise risk for surrogacy arrangements, and should include the welfare of the child.

- New legislation would be necessary to implement the conclusions of the report.
- Surrogate agencies not registered would commit a criminal offence.

The team was not asked to consider two main issues, one of which is pertinent to the success or failure of the surrogacy process:

- No consideration was to be given to the commercialisation of surrogacy.
- No consideration was to be given to enforceability of contracts.

This second point is particularly crucial. The report used some unusual cases of surrogacy reported in the media just prior to the start of the enquiry to inform their debate. The cases referred to consist of a breakdown in communications between a British surrogate mother and her foreign couple, leading to non-relinquishment of the child; a mother carrying a baby for her daughter; a daughter carrying a baby for her mother and a case of a triplet surrogate birth. They also cited unusual American cases of surrogacy including: parents arranging for a surrogate to carry a baby using their deceased daughter's fertilised eggs; and an American director of a commercial company offering surrogacy, arriving in the United Kingdom to tell the British public how for 30,000 pounds they too could have a surrogate baby in the States. Other sources of data used to inform the report were records relating to POs; records kept by guardians ad litem; figures kept by the HFEA; and information from individuals and organisations with an interest in surrogacy, professionals and the public.

The results of a questionnaire consultation of 369 responses were as follows:

1. There should not be a ban on all payments to surrogate mothers.
2. If payment is made, loss of earnings should also be compensated for.
3. Legislation should define the (maximum) amount of remuneration or relevant expenses.

4. A regulatory body should be set up.
5. The law should assist, not restrict, agencies.
6. It should be up to the individual to decide whether or not to use the services of a surrogate agency.

Payment was a considerable focus of the report, where the main fears rested upon baby buying/selling, financial exploitation of a surrogate mother and the exploitation of a surrogate to the commissioning couple through operating in a possibly ever more 'lucrative job'. Furthermore, if surrogacy was accepted as a commercial enterprise it would be contractually enforceable by law, which the team considers an unacceptable outcome. At the same time, unregulated surrogacy as it is today is also unacceptable particularly as the unofficial determination of varied payments is likely to increase disputes regarding finance in surrogacy and add to its perils. The surrogates were at risk as they tend to be of lower socioeconomic positions than commissioning couples and therefore vulnerable. They need legal protection, as do commissioning couples who are vulnerable in their emotional investment of a desire for a child. Thus, the report argued for legislation protecting principally the welfare of the child, the surrogate and the commissioning couple. Other important recommendations made include the knowledge that, as some adverse consequences of surrogate arrangements have been published in the media, the recommendations made a strong case for regulations to attempt to minimise harm, and to monitor the outcome of surrogacy closely in the future so that accurate records of the effects of surrogacy are available. This has not yet been done.

As the general idea was that surrogacy should not under any condition be commercialised, the Brazier Report has given this a lot of attention. Surrogates had been receiving sums of money ranging from less than £100 to £15,000+, although some had carried a baby for others for free. At the moment, it is not unlawful to receive money for surrogacy even if payment is made over and above demonstrable expenses, but there are fears it may be seen as a profession, particularly with some surrogates carrying babies for several couples over many years. According to the HFEA Act 1990 (Section 30), and the 1976 Adoption act, these additional payments could prevent a commissioning couple from obtaining a PO. In practice, although the courts have been made aware of this, no one application has been refused on these grounds.

Section 5.25 of the Report explicitly states what allowable expenses should consist of:

- Maternity clothing
- Healthy food
- Domestic help

- Counselling fees
- Legal fees
- Life and disability insurance

- Travel to and from hospital/clinic
- Telephone and postal expenses
- Overnight accommodation
- Child care to attend hospital/clinic

- Medical expenses
- Ovulation and pregnancy tests
- Insemination and IVF costs
- Medicines and vitamins

It is stressed that documentary evidence of expenses actually incurred be produced by the surrogate mother. In addition to the list provided above, the Report also recommends that an employed surrogate should be reimbursed for loss of actual (not potential) earnings if these were also incurred and documentary evidence of this is provided. In practice these recommendations are not implemented.

Summary

Surrogate motherhood involves numerous medical, social and psychological aspects and risks that are relevant to debates about ethics, contracts and/or regulation. A surrogate mother provides epigenetic contributions to the foetus in vivo in numerous ways; the social and/or genetic parents provide the social epigenetic development of the baby once they take on the responsibility for their development ex utero. In addition in many cases, medical and health care professionals are involved in the bringing about of surrogate pregnancies via assisted conception, be it IVF/ICSI embryo transfers or intrauterine insemination. A surrogate arrangement is therefore neither a purely social arrangement nor an exclusively medical or legal arrangement. Clearly, the legal profession could not solely be responsible for the implementation of contracts if they were legally binding, as these arrangements involve socioemotional 'transactions' as well as medical and financial transactions. Similarly, medical practitioners could not be solely responsible for the contractual arrangements because it can involve sociopsychological changes and legal issues suggesting humanistic values and multidisciplinary involvement must be applied to regulation. Indeed, court practice has shown that in the United Kingdom and the United States, amongst others, judges have overruled the laws of the land and bestowed legal parenthood status to commissioning parents, that is, not to birth parents.

Part III has shown that surrogacy is increasing across the world, can be exploitative (particularly for babies) and prey upon the vulnerability of commissioning couples and surrogates alike. It therefore needs regulating at some level, as was shown in Chaps. 8, 9 and 10. The parties involved in surrogacy mostly enter the agreement to fulfil a need, a need for the commissioning parent(s) to build a family and a belief of the surrogate mother to help them achieve this goal and a need to be compensated for their efforts to bear and deliver a baby for the commissioning couple. Chapter 8 has shown that in cross-border surrogacy the concerns about exploitation and commodification are real. Surrogate motherhood involves numerous medical, social and psychological aspects, and risks that are relevant to debates about moral, ethical and human rights issues are demonstrated in Chap. 9. Dignity of the commissioning and surrogate as well as the relevant 'brokering' parties involved should be a prerequisite alongside the welfare and human rights of the child. Part III has shown that the welfare of the child is not paramount across the world and this needs to be the focus of any future debate, research, policy and practice.

References

Adoption Act. (1958). Legislation.Gov.UK. http://www.educationengland.org.uk/documents/acts/1958-children-act.html. Accessed 19 Oct 2016.

Adoption Act. (1976). Legislation.Gov.UK. http://www.legislation.gov.uk/ukpga/1976/36/contents. Accessed 19 Oct 2016.

Andrews, L., & Douglas, L. (1991). Alternative reproduction. *Southern Californian Law Review, 65*, 1991–1992.

BICA. (2016). http://bica.net/. Accessed 19 Oct 2016.

Blake, L., Casey, P., Readings, J., Jadva, V., & Golombok, S. (2010). 'Daddy ran out of tadpoles': How parents tell their children that they are donor conceived, and what their 7-year-olds understand. *Human Reproduction, 25*, 2527–2534.

Blyth, E., & Hunt, J. (1994). A history of infertility counselling in the United Kingdom. In S. Jennings (Ed.), *Infertility counselling* (pp. 175–190). Oxford: Basil Blackwell.

Blyth, E., Crawshaw, M., Frith, L., & van den Akker, O. (2017). Gamete donors' reasons for and expectations and experiences of, registration with a voluntary donor linking register. *Human Fertility*, 1–11. doi:10.1080/14647273.2017.1292005.

Bowcott, O. (2013, September 26). Intended and birth mother in surrogacy entitled to maternity leave says ECJ. European court of justice legal opinion requires the two women to split the paid leave, each taking at least two weeks. *The Guardian*.

Brazier, M., Campbell, A., & Golombok, S. (1998). *Surrogacy review for health ministers of current arrangements for payments and regulation*. Report of the review team. *Cm 4068*. London: Department of Health.

British Medical Association. (1990). *Surrogacy: Ethical considerations*. London: BMA.

British Medical Association. (1996). *Changing conceptions of motherhood. The practice of surrogacy in Britain*. London: BMA.

Cafcass. (2010). Cafcass guidance for parental orders. http://www.cafcass.gov.uk/pdf/(Oct%2010)%20Final%20Parental%20Order%20Guidance%20May2010.pdf

Californian Family Code. (2013). 7962(3)(f)(2). California Code Family Code. FAM DIVISION 11. MINORS PART 3. CONTRACTS. http://law.justia.com/codes/california/2013/code-fam/division-11/part-3/chapter-3/section-6751. Accessed 19 Oct 2016.

Crawshaw, M., Blyth, E., & van den Akker, O. (2012a). The changing profile of surrogacy in the UK – Implications for policy and practice. *Journal of Social Welfare and Family Law, 34* (3) 1–11.

Crawshaw, M., Purewal, S., & van den Akker, O. (2012b). Working at the margins: The views and experiences of court social workers on parental orders' work in surrogacy arrangements. *British Journal of Social Work.* doi:10.1093/bjsw/bcs045.

Department of Health. (2010). Impact assessment of the Human Fertilisation and Embryology (Parental Order) Regulation 2010. http://www.dh.gov.uk/prod_consum_dh/groups/dh_digitalassets/documents/digitalasset/dh_116502.pdf

Donchin, A. (2010). Reproductive tourism and the quest for global gender justice. *Bioethics, 24*(7), 323–332.

Egliston, K. A., McMahon, C., & Austin, M. (2007). Stress in pregnancy and infant HPA axis function: Conceptual and methodological issues relating to the use of salivary cortisol as an outcome measure. *Psychoneuroendocrinology, 32,* 1–13.

EpiHealth. (2016). http://www.epihealthnet.org/. Accessed 12 May 2016.

Family Code of Ukraine. (2008). Clause 123 and Order of the Health Ministry of Ukraine (Surrogacy Article 7) 23 Dec. (2008). Determination of Origin of the Child Born as a Result of Auxiliary Reproductive Technologies Code 123 Part 2 included in the Family Code of Ukraine 22 Dec 2006, as amended on 9 Sept 2011.

Freeman, T. (2015). Gamete donation, information sharing and the best interests of the child: An overview of the psychosocial evidence. *Monash Bioethics Review, 33*(1), 45–63.

Gardner, D. K., & Lane, M. (2004). *Ex vivo* early embryo development and effects on gene expression and imprinting. *Reproduction, Fertility and Development, 17*(3), 361–370.

Glover, L. (1989) Ethics of new reproductive technologies: The Glover report to the European Commission. https://searchworks.stanford.edu/view/1804475. Accessed 19 Oct 2016.

Harper, J., Kennett, D., & Reisel, D. (2016). The end of donor anonymity: How genetic testing is likely to drive anonymous gamete donation out of business. *Human Reproduction, 31*(6), 1135–1140.

HCCH. (1993). 33: Convention of 29 May 1993 on Protection of Children and Co-operation in Respect of Intercountry Adoption. https://www.hcch.net/en/instruments/conventions/full-text/?cid=69. Accessed 19 Oct 2016.

HCCH Hague Conference on Private International Law. (2012). Overview on the World Organisation for cross-border Co-operation in Civil and Commercial Matters. http://www.hcch.net/index_en.pho?act=text. displaye&tid=26. Accessed 22 May 2016.

HFEA. (2008). HFEA Act, Section 54. http://www.legislation.gov.uk/uksi/2009/2232/made. Accessed 19 Oct 2016.

HFEA. (2013). Code of Practice. http://www.hfea.gov.uk/fertility-treatment-options-surrogacy.html. Accessed 17 Jan 2016.

Hodson, N., & Bewley, S. (2017). Parental orders and the rights of surrogate mothers. *British Journal of Obstetrics and Gynaecology.* doi:10.1111/1471-0528.14565.

Human Fertilisation and Embryology Act. (1990). Legislation.Gov.UK. National Archives. http://www.legislation.gov.uk/ukpga/1990/37/contents. Accessed 22 May 2016.

Human Fertilisation and Embryology Authority. (1998). Legislation.Gov.UK. http://www.hfea.gov.uk/3478.html. Accessed 19 Oct 2016.

International Health Federation. (2005). http://www.who.int/ihr/9789241596664/en/. Accessed 19 Oct 2016.

Jadva, V., Freeman, T., Kramer, W., & Golombok, S. (2009). The experiences of adolescents and adults conceived by sperm donation: Comparisons by age of disclosure and family type. *Human Reproduction, 24*, 1909–1919.

Mac Dougall, K., Becker, G., Scheib, J. E., & Nachtigall, R. D. (2007). Strategies for disclosure: How parents approach telling their children that they were conceived with donor gametes. *Fertility & Sterility, 87*, 524–533.

Maternity Action. (2016). Time off and pay for parents in surrogacy arrangements. https://www.maternityaction.org.uk/advice-2/mums-dads-scenarios/7-adopting-or-involved-in-a-surrogacy-arrangement/time-off-and-pay-for-parents-in-surrogacy-arrangements/

Norton, W., Hudson, N., & Culley, L. (2013). Gay men seeking surrogacy to achieve parenthood. *Reproductive Biomedicine Online, 27*, 271–279.

Ombelet, W., De Sutter, P., Van der Elst, J., & Martens, G. (2005). Multiple gestation and infertility treatment: Registration, reflection and reaction – The Belgian project. *Human Reproduction Update, 11*, 3–14.

Parental Orders. (2008). Human Fertilisation and Embryology Act 2008. http://www.legislation.gov.uk/ukpga/2008/22/pdfs/ukpga_20080022_en.pdf. Accessed 19 Oct 2016.

Parental Orders. (2010). The Human Fertilisation and Embryology (Parental Orders) Regulations 2010. Legislation.Gov.UK. http://www.legislation.gov.uk/ukdsi/2010/9780111491355/contents. Accessed 19 Oct 2016.

Payne, N., & van den Akker, O. (2016). Infertility network UK survey on the impact of fertility problems. Report commissioned by and produced for INUK May 2016.

Purewal, S., Crawshaw, M., & van den Akker, O. (2012). Completing the surrogate motherhood process: Parental order reporters attitudes toward surrogacy arrangements, role ambiguity and role conflict. *Human Fertility, 15*(2), 94–99.

Ramskold, L. H., & Posner, M. P. (2013). Commercial surrogacy: How provisions of monetary remuneration and powers of international law can prevent exploitation of gestational surrogates. *Journal of Medical Ethics, 39*, 397–402.

Surrogacy Arrangements Act. (1985). Legislation.Gov.UK. The National Archives. http://www.legislation.gov.uk/ukpga/1985/49

van den Akker, O. B. A. (2012). *Reproductive health psychology*. Chichester: Wiley-Blackwell. ISBN-13: 978–0470683385.

van den Akker, O. B. A. (2013). For your eyes only: Bio-behavioural and psychosocial research objectives. *Human Fertility, 16*(1), 89–93.

van den Akker, O. (2016). Reproductive health matters. *The Psychologist, 29*(1), 2–5.

van den Akker, O. B. A., Payne, N., & Lewis, S. (2017). Catch 22? Disclosing assisted conception treatment at work. *International Journal of Workplace Health Management*. (Accepted for publication July 2017).

Warnock Report. (1984). *Warnock report*. The report of the committee of inquiry into human fertilisation and embryology. London: HMSO.

Abbreviations

ACOG	American College of Obstetricians and Gynecologists
AID	Artificial Insemination by Donor
AIH	Artificial Insemination by Husband
ALRC	Australian Law Reform Committee
ART	Assisted Reproductive Technology
ASRM	American Society of Reproductive Medicine
BASW	British Association of Social Workers
BICA	British Infertility Counselling Association
BMA	British Medical Association
CEDAW	Convention of the Elimination of Discrimination Against Women
C-section	Caesarean section
Cafcass	Children and Family Court Advisory and Support Service
COTS	Childlessness Overcome Through Surrogacy
CRC	Convention of the Rights of the Child
DI	Donor Insemination
DNA	Deoxyribonucleic Acid
EPQ	Eysenck Personality Inventory
eSET	elective Single Embryo Transfer

© The Author(s) 2017
Olga B.A. van den Akker, *Surrogate Motherhood Families*,
DOI 10.1007/978-3-319-60453-4

ESHRE	European Society of Human Reproduction
ET	Embryo Transfer
FNUK	Fertility Network UK
GRO	General Register Offices
HCCH	Hague Convention of Private International Law
HCIA	Hague Convention for Intercountry Adoption
HFEA	Human Fertilization and Embryology Authority
HIV	Human Immunodeficiency Virus
ICSI	Intracytoplasmic Sperm Injection
IFCOG	International Federation of Gynecology and Obstetrics
IFFS	International Federation of Fertility Societies
ISS	International Social Services
IQ	Intelligence Quotient
IUI	Intrauterine Insemination
IVF	In Vitro Fertilization
LGBT	Lesbian Gay, Bisexual and Transgender
MMPI	Minnesota Multiphasic Personality Inventory
NGDT	National Gamete Donation Trust
NHS	National Health Service
OLRC	Ontario Law Reform Commission
ONS	Office for National Statistics
PGD	Prenatal Genetic Diagnosis
PGS	Prenatal Genetic Screening
PND	Post-Natal Depression
PO	Parental Orders
POR	Parental Order Reporter
PROGAR	Project Group on Assisted Reproduction
PTS	Post-Traumatic Stress
PTSD	Post-Traumatic Stress Disorder
SIDS	Sudden Infant Death Syndrome
SPC	Surrogate Parenting Center
UDHR	Universal Declaration of Human Rights
UK	United Kingdom
UN	United Nations

UNESCO	United Nations Educational, Scientific and Cultural Organization
US	United States
USA	United States of America
WHO	World Health Organization

Index

A

abandoned, 161
abnormalities, 177, 216
abortion, 42, 88, 98, 124, 216, 280
abuse, 138
accredited, 69
accurate birth, 207
accurate record keeping, 209
adjustment problems, 171
adolescence, 155, 203
adoptees, 182
adoption, 85, 101, 120, 182, 234, 286
Adoption Act, 273, 279, 299
Adoption Contact Register, 274
adoptions, 199
adulthood, 155
adverse outcomes, 175
advertising, 67
agency, 92
aggression, 152
alcohol, 82

alienation, 48
altruism, 9, 21, 26, 88, 205, 283
　non-commercial, 84
　surrogacy, 43
American College of Obstetricians and Gynecologists (ACOG), 30
American Society of Reproductive Medicine (ASRM), 16, 30
amniocentesis, 42, 95
anonymity, 29
anonymous, 83, 151
　donor, 210
　surrogacy, 180
ante-natal, 246, 280
anthropologist, 90
anthropology, 47
anxiety, 128, 152, 250, 259
anxious, 154
appeals, 235
appraisal, 155
artificial insemination, 269, 288

© The Author(s) 2017
Olga B.A. van den Akker, *Surrogate Motherhood Families*,
DOI 10.1007/978-3-319-60453-4

atherosclerosis, 176
attachments, 27, 135, 147, 151, 200, 259
attitudes, 50–2, 180, 199, 206–7
autoimmune diseases, 124
autonomous, 45, 51
autonomy, 53, 81, 82, 236, 246

B

baby factories, 218
baby harvesting, 218
Baby Mama, 52
baby selling, 204, 243
behavioural control, 51
beneficence, 81, 236
biogenetic ties, 105
biological bonds, 282
biological relationships, 120
biopower, 233
birth, 215
birth certificates, 207, 213, 294
birth mother, 10, 182, 205
birth record, 249
birth registrations, 131, 182
birthweight, 176
black market, 218, 244
bleeding, 97
bonding, 29, 99, 103, 147, 259, 286
borderline personality disorder, 203
Brazier Report, 297–300
breastfeed, 215, 290
breastfeeding, 127, 160
Brilliant Beginnings, 63
British Association of Social Workers (BASW), 272
British Infertility Counselling Association (BICA), 69, 83, 294–7

British Medical Association (BMA), 20, 272, 284
broker, 185
brokering, 235

C

caesarean sections, 44, 95, 175
cardiovascular risk, 176
celebrity, 86
childbearing, 158
Childlessness Overcome Through Surrogacy (COTS), 83
child molester, 221
child production, 174
children, 23
Children Act, 285
children's rights, 207
child trafficking, 204
Christians, 210
chromosomal, 122
citizenship, 183, 200, 205
citizen status, 202
civil partnership, 292
clinic, 92
Code of Practice, 297
codification, 235
coercion, 175, 237
cognition, 104
cognitive dissonance, 129
cognitive processing, 44
cognitive restructuring, 44, 70, 104, 137, 159
commercial, 279
commercial surrogacy, 18, 43
commercial surrogate offspring, 182
commercialisation, 61, 206
commissioned, 82
commissioning, 8, 14, 119

commissioning couples, 60
commodification, 7, 26, 61, 65, 185, 200, 242–4
commodities, 247, 276
communication, 155, 203
community, 107
compensation, 172
compromised, 82
conception, 106, 215, 237
consensual, 235
consent, 86, 97, 203
consenting, 210, 237
consumerism, 206
contact, 151
contraceptive, 41
contraceptive services, 42
contracts, 8, 62, 162, 213, 232, 243, 298
COTS, 63
counselling, 69, 169, 220, 282, 289
counsellors, 63, 83
courts, 206
criminal, 65, 269
criminal offence, 235
criminal record, 221
criminalisation, 269
criminalised, 235
cross-border, 153, 199, 235
cross-border surrogacy, 22, 184, 246
cruelty, 138
cryopreserved, 238
C-sections, 44, 258
cultural determinism, 47
cultures, 24, 98
custody, 213, 234

D
deceive, 213

deception, 92
denial, 99
depressed, 281
depression, 152, 157, 259
deprivation, 199
desperation, 91
detached, 104
detachment, 90, 160
developed countries, 24
developmental problems, 155
DI, 61, 249
diabetes, 176
diagnostic testing, 247
disabilities, 282
disabled, 30
disabled parents, 67
disclose, 60
disclosure, 131, 178–80, 283
discrimination, 22, 54, 173
dishonesty, 107
disruptive behaviours, 157
dissonance, 105
distress, 102
DNA testing, 134
dominance, 237
donor, 123
donor insemination (DI), 10, 39
donor link register, 208
donors, 80
Down's twin, 85
Down's syndrome, 85, 221
drug, 82
dying, 96

E
egg donors, 234
elective single embryo transfer (eSET), 245

embryonic, 121
embryos, 6, 80, 97
  donations, 234
  loss, 124
  transfers, 123, 288, 301
emotional, 223
emotional distress, 8
emotions, 86
employment, 17
employment rights, 17
empowered, 220, 251
empowerment, 89, 91, 205
endocrine, 100
enforceability, 298
enforceable, 243, 282
enforcement, 269
epigenetics, 50, 172, 173, 247, 294
ethical principles, 81, 199, 232, 237
ethics, 44, 234
ethnographic, 49
European Convention on Human
    Rights, 22
European Society of Human
    Reproduction (ESHRE), 30
expenses, 93, 134, 235, 242
exploitation, 7, 18, 44, 45, 53, 201,
    204–5, 209, 232, 237, 241,
    251, 283, 299
exploitative, 241, 277, 283
exploited, 43
extended family, 27, 93, 127
Eysenck Personality Questionnaire,
    87

F
family, 107
Family Court Advisory and Support
    Service, 286

Family Proceedings Court, 286
fears, 178
femininity, 24
feminist theories, 42–6
feminists, 41
fertility, 125
Fertility Network UK (FNUK), 296
financial, 87
financial gain, 88
financial incentives, 173
foetal, 121, 153
foetal abnormality, 30, 280
foetal alcohol syndrome, 282
Foetal alcohol syndrome, 177
Foetal programming, 247
foetal reduction, 68, 247
foetus, 79, 124
foreign, 235
foster, 155
fragmented, 107
framed, 53
fraud, 174, 214
friends, 52

G

gamete donation, 80, 178–9
gametes, 8, 40, 123
gay, 200
genealogic, 28
general health, 87
General Register Offices (GRO), 131
generational, 119
genetic, 120, 153, 254
genetic conditions, 9
genetic information, 207
genetic link, 28, 70, 104–5, 149,
    282
genetic origins, 294

genetic surrogacy, 5, 8, 153
genetic transmission, 82
genetically related, 270, 271
geographical theories, 46–9
gestational, 120
gestational carrier, 17
gestational hypertension, 97
gestational origins, 271, 294
gestational surrogacy, 8, 153
Glover Report, 279–84
government, 20
grandchild, 108
grandchildren, 186
grandmothers, 186
grandparents, 31, 106, 127
grief reactions, 29
growth restrictions, 177
Guardians ad Litem, 285–6
guidelines, 200, 232, 282
guilt, 270

H

haemorrhage, 258
The Hague Conference on Private
    International Law (HCCH),
    184, 273
Hague Convention on inter-country
    surrogacy, 234
health behaviours, 157, 158
health care, 95
health visitor, 94
heterosexual, 52
high blood pressure, 176
High Court order, 273
high risk, 246
HIV, 64
hormonal, 101

hospital records, 207
human dignity, 276
Human Fertilisation and
    Embryology Act (HFEA), 20,
    22, 272, 288–94, 299
Human Fertilization and
    Embryology Authority
    (HFEA), 30
human rights, 201, 237, 253, 294
human trafficking, 209
humanistic, 234
hypertension, 97
hysterectomy, 97
hysterosalpingography, 124
hysteroscopy, 124

I

ICSI, 61
identifiable, 63
identification, 236
identity, 48, 182–3, 208, 253–5
ideology, 49
illegal, 137, 206, 244, 269, 281
illegal drugs, 95
immigration, 221
implications counselling, 69
in vitro fertilisation (IVF), 10, 40,
    61, 84
in vivo, 282, 301
incidence, 19–22
incubators, 236
individuals commissioning a
    surrogate, 119
Industrial Revolution, 90
inequalities, 53, 140, 199, 200, 251
infectious diseases, 9
infertility, 23, 40, 120, 218

information, 92, 199
  counselling, 69
  psychological, 70
informed, 85–6
informed consent, 44, 175, 207, 209
inherit, 254
inheritance, 214, 255
inherited damaged DNA, 170
injustice, 214, 237
insecure, 155
insecurely attached, 155
insemination, 11, 40, 288
intended couples, 119
inter-country adoptions, 173, 174, 179, 207, 256
intergenerational, 106
international adoptions, 184
International Federation of Gynecology and Obstetrics (IFCOG), 30
international laws, 222, 254
international surrogacy, 53, 68
intra-familial, 187
intrauterine, 82
intrauterine insemination (IUI), 12, 301
investment, 171
involuntary childlessness, 24, 120

J
Jews, 210
justice, 81, 221, 232

K
karyotypes, 124
kinships, 39–41, 107, 109, 138–9, 214

known gamete donations, 179

L
legal, 83, 128, 137, 221
  parenthood, 179
  precedent, 201
  status, 286
legislation, 7, 43, 88, 149, 256, 298, 299
legitimising, 139
lesbian, 151
liberty, 237, 269
licence, 288
licensed clinics, 288
life satisfaction, 157
lifestyle, 60
litigation, 97
local authority, 273
loss, 98
low birthweight, 152, 175

M
malformations, 175
malpractice, 173, 201
maltreated, 155
marginalisation, 251
marital, 107
marital relationship, 275
masculinity, 24
maternal, 121
  behaviour, 156
  instinct, 104
  sensitivity, 155
  baby, 127
  foetal, 127, 158
  foetal attachments, 99
  foetal detachment theory, 102

maternity, 40, 94–6, 108, 243
maternity leave, 250, 290, 291
mazing, 130
media, 86, 104
medical ethics, 236
medical morbidity, 245
medical tourism, 204
mental health, 60, 155, 252
microcephaly, 177
midwifery, 94
Minnesota Multiphasic Personality
    Inventory (MMPI), 87
miscarriages, 30, 88, 98, 123, 152,
    216
misleading, 89
mitochondria, 10
mitochondrial donation, 172
monitored, 176
monitoring, 220, 247–8
moral, 18, 44, 232
morality, 200, 234
morally ambiguous, 206
mortality, 245
mother, 17
mother-child, 276
mother-child bonding, 27
motherhood, 211
motivations, 40, 51, 88
multidisciplinary, 172
multiparental, 161
multiple births, 152, 245
multiples, 220
Muslim, 214

N

national surrogacy, 22
nationality, 253, 272
negative experiences, 92

neglect, 138
negligence, 249
neonatal, 82
    death, 30, 98, 107, 124
    outcomes, 97
NHS, 275
non-commercial, 88, 283
non-maleficence, 81, 236
non-relinquishment, 270
non-traditional, 40, 137, 283
normative beliefs, 51
nurture, 185
nurturing, 105
nutritional, 60

O

objectification, 45
obstetric, 96–8
offspring, 80
oocyte, 221
    donation, 10, 175
    donor, 97
    retrieval, 12, 125
openness, 209
origins, 213, 283
ostracisation, 92
overprotective, 152

P

paedophilia, 138
Parent Responsibility Agreement,
    279
parentage, 200
parental custody, 235
parental fitness, 65
parental orders, 22, 286–7
parental rights, 235, 244

parental stress, 152
parenthood, 211, 272
parenting, 23, 39, 147, 152
parenting stress, 152
parents, 6
partner, 93
passport, 210
paternal attachment, 152
paternity, 40, 108
patriarchal, 233
payment, 235
pensions, 294
perinatal, 175, 247
perinatal loss, 124
personality, 87
    disordered, 155
    profiles, 87
placenta, 246
placenta accrete, 97
placental disorders, 97
policies, 235
political feminist models, 41–2
politics, 46
POs, 131
positive affectivity, 157
positive attitudes, 159
post relinquishment, 103
post-delivery, 160
post-natal, 99, 216, 246
post-natal attachment, 99
post-natal depression (PND), 82, 93,
    107, 257, 259
post-natal psychosis, 257
post-partum, 95, 96, 246, 257
post-relinquishment, 103
post-traumatic stress (PTS), 98
post-traumatic stress disorder
    (PTSD), 98

poverty, 173, 237
power, 233
powerlessness, 251
pre-birth contracts, 293
pre-birth orders, 85
pre-birth PO, 255
precariat, 48
precarity, 48
preconception, 95, 149
preconceptual testing, 64
pre-eclampsia, 97
pregnancy, 16, 120, 159, 216, 247
premature birth, 175
prematurity, 152
prenatal, 99
    attachment, 99, 136
    care, 95
    genetic diagnosis, 50, 122
    screening, 247
    testing, 64, 216
previous losses, 88
procreation, 241, 242
procreative liberty, 240–1
professionals, 105
prohibitory, 235
Project Group on Assisted
    Reproduction (PROGAR),
    272
pro-life, 210
prostitution, 215
psychiatric, 82, 87
psychiatric screening, 82
psychology, 92, 185, 202, 216, 245
    assessments, 60, 96, 171
    damage, 276
    distress, 103, 119, 248
    effects, 270
    hardship, 259

morbidity, 245
needs, 70
profiles, 87
risks, 256
state, 122
testing, 62
welfare, 294
wellbeing, 103
compromised, 270
psychopathology, 87, 151
psychosexual counselling, 70
psychosis, 259
psychosocial, 161
psychosocial development, 152
psychosocial screening, 61

Q
qualitative, 90
quality of life, 87

R
reasonable expenses, 88
record keeping, 179, 236, 247
records, 95
registered, 208
registration, 215
regulating, 288
regulations, 9, 61, 200, 232, 234,
    271, 279
regulatory, 235
relatedness, 105
relationship, 92
religion, 210, 214, 216
relinquishment, 0, 68, 99, 102–4,
    159, 216, 220, 246, 270, 282,
    283

renege, 68
reporting, 247
reproduction
    autonomy, 240
    freedom, 44
    loss, 124
    tourism, 48
repro-travellers, 219
restructuring, 104
risks, 256
role ambiguity, 287
Roman Catholic, 50, 210

S
same-sex, 5, 210, 235
schizophrenia, 173
screening, 282
secrecy, 132, 182
secure attachment, 155
securely attached, 155
selective feticide, 42
self-confidence, 157
self-efficacy, 152
self-esteem, 89, 150, 157
self-interest, 89
self-worth, 205
sensitive period, 100
separation, 39, 147, 155
separation anxiety, 156
sex selection, 68, 222, 223
sex selective abortions, 68
sexual exploitation, 218
Shiite, 214
single, 151, 200, 235
single men, 120
single parents, 292
single women, 22

smoking, 82
social cognition, 154
social geometry, 47
social mobility, 47
social motherhood, 105
social parents, 14
social relations, 47
social stigma, 93, 134
social support, 28–9, 96, 106, 259
sociocultural, 148
sociocultural context, 6
sociodemographic, 233
socioeconomic, 23, 91, 211
socioemotional adjustment, 152
socioemotional risk, 270
sociopsychological, 233
sperm, 178
    count, 64
    donation, 24
    donors, 234
stateless, 183, 206, 253
stigma, 40, 81, 178, 218, 259
stigmat, 23
stigmatising, 107, 215
stillbirth, 30, 98, 124
stressful, 157
structuralism-functionalist family
    theory, 40–1
sub-fertile, 248
subjective wellbeing, 157
succession rights, 294
sudden infant death syndrome
    (SIDS), 93
suicidal, 297
Sunni, 214
surplus embryos, 80
Surrogacy Act, 83, 84

Surrogacy Arrangements Act, 20,
    235, 275, 278–9
surrogacy UK, 63
surrogate agencies, 298
surrogate births, 95
surrogate motherhood, 59
symbolic, 239
symbolism, 47

T

taxation, 294
temperament, 161
termination, 88, 95
theory, 40, 129
theory of planned behaviour, 51
third-party, 120, 125, 276
third-party reproduction, 53
third-party-assisted conception, 6
traditional, 13, 211
traditional attitudes, 40
traditional family, 40
traffic children, 185
transfer of legal parenthood, 253
trans-placental, 294
transracial adoption, 182
traumatised, 103
trimester, 159
triplet, 186, 298
trust, 232
twins, 84, 97, 220, 245
twins, 220

U

ultimately, the power here lies with the
    surrogate, and the lack of, 214

unbiased, 86
unenforceable, 62, 277, 281
unintended disclosure, 179
The Universal Declaration of Human
     Rights, 231
The Universal Declaration on
     Bioethics and Human Rights
     (UNESCO), 231
unregulated, 237
uterine rupture, 97

V

vasa praevia, 246

violence, 251
vulnerable, 173, 251

W

Warnock Committee, 275
Warnock Report, 275
welfare, 173, 202–6
the welfare of the child, 150, 184–7,
     239
wellbeing, 157
wet nurse, 214
World Health Organization (WHO),
     273

The manufacturer's authorised representative in the EU is Springer
Nature Customer Service Centre GmbH, Europaplatz 3, 69115 Heidelberg,
Germany. If you have any concerns regarding our products, please
contact ProductSafety@springernature.com

Printed and bound by CPI Group (UK) Ltd, Croydon, CR0 4YY
27/04/2026
02097621-0001